Home Food Preservation

HE-1

Evelyn F. Crayton
Extension Assistant Director for Family and Community Programs
Professor, Nutrition and Food Science
Auburn University

2006

ALABAMA COOPERATIVE **Extension** SYSTEM
Your Experts for Life

Alabama A&M and Auburn Universities

ALABAMA
COOPERATIVE
Extension
S Y S T E M
Your Experts for Life

www.aces.edu

Issued in furtherance of Cooperative Extension work in agriculture and home economics, Acts of May 8 and June 30, 1914, and other related acts, in cooperation with the U.S. Department of Agriculture. The Alabama Cooperative Extension System (Alabama A&M University and Auburn University) offers educational programs, materials, and equal opportunity employment to all people without regard to race, color, national origin, religion, sex, age, veteran status, or disability.

© 2006 by the Alabama Cooperative Extension System. All rights reserved.

ISBN-10: 0-9722580-8-6
ISBN-13: 978-0-9722580-8-1

Table of Contents

Food Preservation 5

Canning 16
Food Safety and Quality 16
General Steps in Canning 19
Materials and Equipment 21
Methods and Procedures 30
Canning Problems 35
Canning Fruits 39
Canning Tomatoes and Tomato Products 59
Canning Vegetables 71
Canning Meat, Poultry, and Seafood 87
Canning Nuts 93

Making Pickles and Other Pickled Products 94
Ingredients 94
Methods and Procedures 96
Materials and Equipment 98
Common Causes of Problems 99
Pickle Recipes 100

Making Jelly and Other Jellied Spreads 124
Ingredients 124
General Steps in Making Jelly 126
Materials and Equipment 130
Common Causes of Problems 130
Jelly Recipes 132
Making Other Jellied Spreads 138
Jam Recipes 139
Preserves Recipes 143
Marmalade Recipes 148
Conserve Recipes 152
Fruit Butter Recipes 155
Fruit Syrup Recipe 161
Refrigerated Spread Recipes 162

Canning for Fairs and Exhibits 163

Freezing 168
Food Safety and Quality 168
General Steps in Freezing 169
Materials and Equipment 170
Methods and Procedures 175
Common Causes of Problems 181
Freezing Fruits 183
Freezing Vegetables 194
Freezing Meat, Poultry, and Seafood 203
Freezing Eggs 209
Freezing Nuts 210
Freezing Dairy Foods 211
Freezing Cooked Foods 212

Drying 217
Food Safety and Quality 218
General Steps in Drying 218
Materials and Equipment 219
Methods and Procedures 222
Fruit Recipes 230
Vegetable Recipes 233
Making Fruit and Vegetable Leathers 237
General Steps in Making Leathers 238
Puree Recipes for Leathers 239
Drying Herbs 242
Drying Peanuts, Popcorn, and Seeds 244

Timetables 246

References 252

Index 253

List of Tables

Title	Page
Table 1. Critical Temperatures for Food Preservation.	10
Table 2. Acid Level (pH) in Foods Likely To Be Home Preserved.	11
Table 3. Boiling Point at Various Altitudes.	11
Table 4. Approximate Number of Fresh Items in One Pound or Pint.	12
Table 5. Number of Pounds, Quarts, and Pints in a Bushel of Fruits.	13
Table 6. Number of Pounds, Quarts, and Pints in a Bushel of Vegetables.	14
Table 7. Ideal Weight for Slaughter and Pounds of Usable Meat.	14
Table 8. Measuring Equivalents.	15
Table 9. Syrups for Canning.	40
Table 10. Pounds of Fresh Fruits Needed for Standard Canner Loads.	41
Table 11. Pounds of Fresh Vegetables Needed for Standard Canner Loads.	72
Table 12. Extracting Juice: Cups of Water Per Pound of Fruit.	126
Table 13. Jellying Point at Critical Altitude Differences.	128
Table 14. Headspace to Allow in Freezer Packages.	179
Table 15. Recommended Storage Times for Home-Frozen Foods.	180
Table 16. Foods not Suitable for Freezing.	182
Table 17. Syrups for Freezing.	185
Table 18. Pounds and Pints of Dried Foods From Fresh.	217
Table 19. Suitability of Vegetables for Drying.	222
Table 20. Suitability of Fruits for Drying.	223
Table 21. Suitability of Foods for Making Leathers.	237
Table 22. Canning Timetable for High-Acid Fruits.	246
Table 23. Canning Timetable for Low-Acid Vegetables.	247
Table 24. Canning Timetable for Meats, Poultry, and Fish.	247
Table 25. Blanching Timetable for Freezing Vegetables.	248
Table 26. Blanching and Drying Timetable for Fruits.	249
Table 27. Blanching and Drying Timetable for Vegetables.	250

Food Preservation

Canning, freezing, and drying allow you to preserve the flavor and nutrients of fresh food so that you may enjoy and benefit from them long after they are in season. The appearance, flavor, texture, and nutritional value will be altered to some extent, but when properly preserved, the food will be good for you and will taste good, too.

Foods normally begin to spoil very shortly after they are harvested. Spoilage results from chemical changes such as those caused by enzymes, from growth of microorganisms such as molds and bacteria, and from physical damage such as water loss and bruising. Food preservation methods do destroy some microorganisms, but primarily they only stop the spoilage process during the time that the food is sealed and stored. As soon as the canning jar or freezer package is opened, the foods begin to deteriorate again.

There are three major categories of food preservation: canning, freezing, and drying. Pickles and jellied products must be canned to be preserved for an extended time. But, they are treated separately in this book because the procedures for preparing pickles and jellies before canning them are so different from preparing fruits and vegetables that will simply be canned.

The following Glossary Of Terms will be helpful if you are preserving food for the first time. Beyond the Glossary are several tables that should be helpful, as well. And, the experienced food preservers will find the timetables near the end of the book a handy reference.

GLOSSARY OF TERMS

Acid Foods
Foods that contain enough acid to result in a pH of 4.6 or lower. Bacteria do not grow well in acid foods; consequently, they may be processed in a boiling water-bath canner. Acid foods include all fruits except figs, most tomatoes, fermented and pickled vegetables, and jellied spreads.

Acidification
In food preservation, a method of adding acid to low-acid foods. Lemon juice, vinegar, and fruit juices are included in many recipes to add acid. Also, citric acid products can be bought in the canning section of most grocery stores to acidify low-acid foods.

Altitude
The vertical elevation of a location above sea level. Altitudes higher than 2000 feet above sea level affect the boiling point of water. This, in turn, can affect the safety of canned foods, and processing times must be lengthened. However, no location in Alabama is higher than 1000 feet.

Ascorbic Acid
The chemical name for vitamin C. Lemon juice contains large quantities of ascorbic acid and is commonly used to prevent browning of peeled, light-colored fruits and vegetables. Ascorbic acid, ascorbic acid powder, and ascorbic acid mixture can be bought for use in food preservation.

Bacteria
A large group of one-celled microorganisms that occur naturally. See Microorganisms (p. 8).

Blancher
A 6- to 8-quart pot with a lid and a fitted perforated basket to hold food in boiling water (for water blanching) or a fitted rack to hold food over water (for steam blanching). It is also useful for loosening skins on fruits to be peeled or for heating foods to be hot packed. If you don't have a blancher, you can use any large pot with a tight-fitting lid with a metal colander, sieve, deep-fry basket, or cheesecloth bag that will fit inside. For steaming, you will need a metal rack that will hold the food 3 inches above 1 to 2 inches of boiling water.

Boiling Water-Bath Canner
A large standard-size pot with a lid and a jar rack designed for processing 7 quarts or 8 to 9 pints of food in boiling water. If you don't have a canner, you can use any large, heavy pot with a tight-fitting lid and a metal rack that will fit inside to hold the jars off the bottom of the pot.

Botulism
An illness caused by eating a toxin (poison) produced by growth of *Clostridium botulinum* bacteria. This bacterium grows in moist, low-acid food that contains less than 2 percent oxygen and is stored between 40° and 120°F. Proper heat processing destroys the bacteria when canning food. Freezer temperatures do not destroy it but do prevent its growth in frozen food. A low moisture level controls its growth in dried food, and a high oxygen level controls its growth in fresh food.

Canning
A method of preserving food. Foods are placed in jars or cans and heated to a temperature that destroys microorganisms and inactivates enzymes. This heating and later cooling forms a vacuum seal which prevents other microorganisms from contaminating the food while it is sealed.

Canning Salt
A purer salt than table salt. Also called pickling salt, it does not have iodine or the anti-caking additives which can discolor canned foods and make them look cloudy.

Citric Acid
A form of acid that can be added to foods when canning. It increases the acidity of low-acid foods and may improve the flavor and color.

Cold Pack
A canning procedure in which jars are filled with raw food. **Raw pack** is another name for this procedure and is the more preferred term. **Cold pack** is sometimes used incorrectly to refer to foods that are canned in a pot with no lid or to jars of food that are processed in boiling water.

Enzymes
Proteins in food which accelerate many flavor, color, texture, and nutritional changes, especially when food is cut, sliced, crushed, bruised, and exposed to air. These changes are not always bad; for example, the aging of meat. But, enzymes also cause peeled fruit to turn dark and food tissue to become mushy. Proper blanching or hot packing practices destroy enzymes and improve food quality.

Exhausting
Removing air from within and around food and from jars and canners. Blanching exhausts air from food tissues. Exhausting, or venting, a pressure canner is necessary to prevent the risk of *Clostridium botulinum* developing in low-acid canned foods.

Fermentation

Changes in food caused by intentional growth of bacteria, yeast, or mold. Native bacteria ferment natural sugars to lactic acid, a major flavoring and preservative in sauerkraut and in naturally fermented dill pickles. Alcohol, vinegar, and some dairy products are also fermented foods.

Headspace

The unfilled space above food or liquid in jars or in freezer containers. For canning, headspace allows for food expansion as jars are heated and for forming vacuums as jars cool. In freezing, headspace allows for food expansion as it freezes.

Heat Processing

Treating jars of food with enough heat to allow safe storage at normal home temperatures.

Hot Pack

A canning procedure in which raw food is heated in boiling water or steam and then placed into hot jars.

Low-Acid Foods

Foods which contain very little acid and have a pH above 4.6. The acidity in these foods is not sufficient to prevent the growth of the bacterium *Clostridium botulinum.* Vegetables, some tomatoes, figs, all meats, poultry, seafoods, and some dairy foods are low acid. To control the risk of botulism, jars of these foods must be heat processed in a pressure canner or acidified to a pH of 4.6 or lower before processing in a boiling water-bath canner.

Microorganisms

Independent organisms of microscopic size, including bacteria, yeast, and mold. When alive in a suitable environment, they grow rapidly and may divide or reproduce every 10 to 30 minutes. Therefore, they reach high populations very quickly. Undesirable microorganisms cause disease and food spoilage. Microorganisms are sometimes intentionally added to ferment foods, make antibiotics, and for other reasons.

Mold

A fungus-type microorganism whose growth on food is usually visible and colorful. Molds may grow on many foods, including acid foods like jams and jellies and canned fruits. Using the recommended heat processing and sealing practices prevent their growth on these foods.

Mycotoxins

Toxins (poisons) produced by the growth of some molds on foods.

Pasteurizing
Heating a specific food enough to destroy the most heat-resistant pathogenic or disease-causing microorganism known to be associated with that food.

pH
A measure of acidity or alkalinity. Values range from 0 to 14. A food is neutral when its pH is 7.0; lower values are increasingly more acid; higher values are increasingly more alkaline.

Pickling
The procedure of adding enough vinegar or lemon juice to a low-acid food to lower its pH to below 4.6. Properly pickled foods may be safely heat processed in a boiling water-bath canner.

Pressure Canner
A specially designed metal pot with a lockable lid used for heat processing low-acid food. These canners have jar racks, one or more safety devices, systems for exhausting air, and a way to measure or control pressure. Canners commonly have a 20- to 21-quart capacity. The smallest canner that can be safely used has a 16-quart capacity and holds 7 quart jars. Pressure saucepans with less than 16-quart capacities are not recommended for canning.

Raw Pack
A canning procedure in which jars are filled with raw, unheated food. It is acceptable for canning low-acid foods in a pressure canner, but it allows more rapid quality loss in acid foods that are heat processed in a boiling water bath.

Spice Bag
A closeable fabric bag used to hold spices. Placed in the pickling solution and then moved before packing jars, it prevents pickles from darkening.

Style of Pack
Form of canned food, such as whole, sliced, juice, or sauce. The term is also used to identify whether food is packed raw or hot into jars. For frozen foods, styles of packaging include dry, syrup, sugar, liquid, and tray packs.

Vacuum
The state of negative pressure. It reflects how thoroughly air is removed from within a jar of processed food: the higher the vacuum, the less air left in the jar.

Yeasts
A group of microorganisms which reproduce by budding. They are used in fermenting some foods and in leavening breads.

FOOD PRESERVATION TABLES

Table 1. Critical Temperatures for Food Preservation

Canning temperatures are used to destroy most bacteria, yeasts, and molds in foods. Time required to kill these decreases as temperatures increase.	250°F	Canning temperatures for low-acid vegetables, meat, and poultry in a pressure canner.
	240°F	
Temperature at which water boils at sea level.	212°F	Canning temperature for acid fruits, tomatoes, pickles, and jelly products in a boiling water-bath canner at sea level.
	180°F	
	165°F	
Temperatures at which foods are warmed may prevent growth of microorganisms but may allow some to survive.	140°F	
DANGER ZONE Temperatures between 40°-140°F allow rapid growth of bacteria, yeasts, and molds.	95°F	Maximum storage temperature for canned foods.
	70°F	
		Best storage temperatures for canned and dried foods.
	50°F	
	40°F	
Temperature at which water freezes at sea level.	32°F	
Cold temperatures permit slow growth of some bacteria, yeasts, and molds.		
Freezing temperatures stop growth of microorganisms, but may allow some to survive.		
	0°F	Best storage temperatures for frozen foods.
	−10°F	

Table 2. Acid Level (pH) in Foods Likely to be Home Preserved

pH	Category	Foods
1.0 – 3.0	Strong Acid	
		plums
		dill pickles, apricots
		apples, blackberries, gooseberries
		sour cherries, peaches
		sauerkraut, raspberries
		strawberries, blueberries
		sweet cherries
		pears
4.0		
		most tomatoes
4.6	High Acid	
	Low Acid	
		okra
5.0		squash, pumpkins, carrots
		pimientos, turnips
		green peppers, beets, snap beans
		sweet potatoes
		Irish potatoes, spinach, asparagus
		mustard greens, figs, baked beans
		red kidney beans
		lima beans
6.0		
		chicken, succotash
		peas
		corn
		mushrooms
		most meats
7.0	Neutral	hominy, ripe olives
14.0	Strong Alkali	

Table 3. Boiling Point at Various Altitudes*

Sea Level	2,000 Feet	4,000 Feet	6,000 Feet	8,000 Feet
212°F	208°F	204°F	201°F	197°F

*No point in Alabama is higher than 1,000 feet above sea level.

Table 4. Approximate Number of Fresh Items in One Pound or Pint

Fruit or Vegetable (Form as Weighed)	Number in One Pound
Fruits	
Apples	3 medium
	3 cups sliced
Apricots	12 medium
Cherries (whole)	about 50
Figs	9 large, 14 small
Grapefruit	1½ medium
Nectarines	9 medium
Oranges	3 medium
Peaches	4 medium
Pears	3 medium
Plums	9 medium
Rhubarb	2 cups cooked
Strawberries	22 medium
Tomatoes	4 medium
Vegetables	
Asparagus	2 cups of 1-inch pieces
Lima Beans (in pods)	1⅛ cups shelled
Snap Beans	3 cups of 1-inch pieces
Broccoli	1⅔ cups of 1-inch pieces
Carrots	2 cups of ½-inch slices
Cauliflower	2 cups of flowerets
Corn (in husks)	3 pieces
	1⅓ cups of whole kernels
Cucumbers	2 large
Mushrooms (whole)	30 medium
Okra	35 medium
Onions	1 large
Garden Peas (in pods)	1 cup shelled
Southern Peas (in pods)	1 cup shelled
Sweet Peppers	2 large
Irish Potatoes	3 medium
Spinach	12 cups
Summer Squash	3 medium
	4 cups sliced
Sweet Potatoes	3 medium
Turnip Roots	3 cups cubed
Miscellaneous	
Nuts, in shells	2 cups nutmeats
Nuts, shelled	3 to 4 cups nutmeats
Raisins	3 cups

Table 5. Number of Pounds, Pints, and Quarts in a Bushel of Fruits

Type and Form of Fruit	Pounds of Whole Fruit in 1 Bushel	Pints and Quarts of Prepared Fruit in 1 Bushel	
		Pints	Quarts
Apples, sliced	50	16-20	8-10
Applesauce	50	15-18	7-9
Apricots, -halved, sliced	50	18-24	9-12
Berries, except Strawberries and Cranberries	36+	12-18	6-9
Cherries, whole	56	22-32	11-16
Cranberries	100		
Grapefruit	40++		
Grapes, whole	48	10-12	5-6
Grape Juice	48		
Nectarines, -halved, sliced	48	16-24	8-12
Peaches, -halved, sliced	48	19-25	9-12
Pears,-halved, quartered	50	20-25	10-12
Pineapple,-sliced, cubed	70+	20-28	10-14
Plums, whole, halved	56	24-30	12-15
Rhubarb, stewed	28+++	12-20	6-10
Strawberries	36+	12-16	6-8
Tomatoes, -whole, halved	53	15-20	7-10
Tomatoes, crushed	53	16-22	8-11
Tomato Juice	53		
Tomato Sauce, thin	53		
thick	53		

+Pounds of whole berries in a 24-quart crate rather than a bushel.
++Pounds of whole fruit in a bag rather than a bushel.
+++Pounds of whole fruit or berries in a lug rather than a bushel.

Table 6. Number of Pounds, Pints, and Quarts in a Bushel of Vegetables

Type and Form of Vegetable	Pounds* of Whole Vegetable in 1 Bushel	Pints and Quarts of Prepared Vegetable in 1 Bushel	
		Pints	Quarts
Asparagus	30+	10-15	5-7
Lima Beans, shelled	32	5-8	2-4
Snap Beans, pieces	30	15-20	7-10
Beets, whole, sliced	52	17-20	8-10
Broccoli	25+	10-12	5-6
Carrots, sliced, diced	50	16-20	8-10
Corn, whole-kernel	35	8-9	4-5
Cucumbers	48	24-30	12-15
Okra, whole, sliced	26	19-21	9-10
Garden Peas	30	6-8	3-4
Southern Peas, shelled	25	6-7	3-4
Hot Peppers	25	17-21	8-10
Sweet Peppers	25	17-21	8-10
Potatoes, whole, cubed	50	18-22	9-11
Spinach	18	4-9	2-4
Winter Squash++, cubed	40	16-20	8-10
Sweet Potatoes, whole	50	16-25	8-12
Turnip Greens	18	8-9	4-5

*Vegetables weighed as follows: beans in shells; beets and carrots without tops; corn in husks; peas in pods.
+Pounds of whole vegetable in 1 crate rather than 1 bushel.
++Do not can summer squash; freeze them.

Table 7. Ideal Weight for Slaughter and Pounds of Usable Meat

Meat	Ideal Live-Weight for Slaughter	If Live-Weight Is:	Approximate Amount of Usable Meat Is:
Beef	800-1,000 lb.	1,000 lb.	425 lb.
Lamb	80-100 lb.	80 lb.	32 lb.
Veal	125-180 lb.	125 lb.	63 lb.
Pork	200-225 lb.	200 lb.	107 lb. + 19 lb. fat

Table 8. Measuring Equivalents

1 dash or pinch	=	2 to 3 drops or less than ⅛ teaspoon
1 jigger	=	1½ fluid ounces
1 tablespoon	=	3 teaspoons
		½ fluid ounce
¼ cup	=	4 tablespoons
		2 fluid ounces
⅓ cup	=	5 tablespoons + 1 teaspoon
½ cup	=	8 tablespoons
		4 fluid ounces
1 cup	=	16 tablespoons
		8 fluid ounces
1 pint	=	2 cups
		16 fluid ounces
1 quart	=	4 cups
		2 pints
		32 fluid ounces
1 gallon	=	4 quarts
		128 fluid ounces
8 solid quarts	=	1 peck
4 pecks	=	1 bushel
16 solid ounces	=	1 pound

Canning

Canning is a safe and economical way to preserve food. Canning home-grown food may save you as much as half the cost of buying commercially canned food, if you already have gardening equipment and supplies and don't count the value of your labor. And, preparing home-canned relishes, jams, jellies, and pickles for family and friends is a fulfilling experience and a source of pride for many people.

Canning helps preserve the nutritive value of fresh fruits and vegetables. Many foods begin to lose some of their vitamins as soon as they are harvested. In fact, nearly half the vitamins may be lost within a few days unless the produce is cooled or preserved immediately. Within 1 to 2 weeks, even refrigerated produce loses half or more of some vitamins.

The heating process during canning does destroy from one-third to one-half of the very sensitive vitamins A and C, thiamin, and riboflavin. Once canned, additional losses of these vitamins are from 5 to 20 percent during each year of storage. The amounts of the other vitamins, however, are only slightly lower in canned foods than in fresh. If vegetables are handled properly and canned promptly after harvest, they can be at least as nutritious as the fresh produce you buy in local supermarkets that has been harvested, shipped, and then handled by customers and market personnel.

FOOD SAFETY AND QUALITY

There are several reasons why fresh foods spoil or lose their quality. Most fresh foods have a high percentage of water, making them very perishable. Microorganisms live and multiply quickly on the surfaces of fresh foods and on the inside of bruised, insect-damaged, and diseased foods. Also, the natural actions of oxygen and the enzymes that are present throughout fresh food tissues can reduce food quality if not treated early.

Proper canning practices remove oxygen from food tissues, destroy enzymes, prevent the growth of undesirable bacteria, yeasts, and molds, and help form a high vacuum in jars. Good vacuums form tight seals which keep liquid in and air and microorganisms out.

Food Safety

The growth of *Clostridium botulinum* in canned food can cause botulism—a deadly form of food poisoning. These bacteria exist either as spores or as vegetative cells. *Botulinum* spores, which are similar to seeds, are on most fresh food surfaces. They can survive harmlessly in soil and water for many years. However, when ideal conditions exist for growth, the spores produce vegetative cells which multiply rapidly. The cells produce the deadly toxin within 3 to 4 days.

Ideal conditions for cell production include:
- a moist, low-acid food
- a temperature between 40° and 120°F
- less than 2 percent oxygen, which occurs in tightly packaged foods

Most bacteria, yeasts, and molds are difficult to remove from food surfaces. Washing fresh food reduces their numbers only slightly. Peeling root crops, underground stem crops, and tomatoes reduces their numbers greatly. Blanching also helps. *But the only sure control is a high acid content in the food or high-temperature processing for the recommended amount of time.*

Using the recommended processing time ensures destruction of heat-resistant microorganisms in home-canned foods. The recommended time is based on the largest number of microorganisms expected to be present on a given food. Home-canned foods will be free of spoilage if they are properly prepared and processed, the lids are sealed to form a high vacuum, and they are stored at temperatures below 95°F (preferably between 50° to 70°F).

High Acid Content. The amount of acid in a food determines whether it should be processed in a pressure canner or water-bath canner. Acid blocks the growth of *Clostridium botulinum* bacteria or destroys them more rapidly when heated. The acid may be natural, as in most fruits, or it may be added, as in pickled foods. Acid foods are not at risk for botulinum growth. Botulinum spores in low-acid foods, however, must be destroyed with high heat.

The term "pH" is a measure of acidity; the lower the pH value, the higher the acid content of the food.

Acid foods have a pH of 4.6 or lower. They include fruits, except figs and some varieties of tomatoes. The acid content of many other foods can be increased by adding lemon juice, citric acid, or vinegar. Canned fruits and foods made from fruits, such as jams, jellies, marmalades, and fruit butters, and foods with added acid, such as pickles and sauerkraut, can be safely canned in a water-bath canner.

Low-acid foods have pH values higher than 4.6. They include red meats, seafood, poultry, milk, all fresh vegetables, and some varieties of tomatoes and must be canned in a pressure canner. Mixing low-acid with acid foods does not lower their pH below 4.6 unless the recipes include enough lemon juice, citric acid, or vinegar to acidify them (make them acid foods).

18 • Canning/General

Although tomatoes have usually been considered an acid food, some varieties are now known to have pH values slightly above 4.6. Therefore, it is safer to consider all tomatoes as a low-acid food. Figs also have pH values slightly above 4.6. Consequently, if they are to be canned in a water-bath canner, lemon juice or citric acid must be added. Properly acidified tomatoes and figs can be safely processed in a water-bath canner.

Botulinum is very hard to destroy at the temperatures that can be reached in a water-bath canner. The higher the canner temperature, the more easily the spores are destroyed. Therefore, all low-acid foods should be processed at temperatures of 240° to 250°F. This temperature range can be reached only with pressure canners operated at 10 to 15 PSIG. (PSIG means pounds per square inch of pressure as measured by a gauge. PSI is the more familiar term and will be used hereafter in this publication.)

At temperatures of 240° to 250°F, the time needed to destroy bacteria in low-acid food ranges from 20 to 100 minutes. The exact time depends on the kind of food being canned, the way it is packed in the jars, and the size of the jars. The time needed to safely process low-acid foods in a water-bath canner ranges from 7 to 11 hours; the time needed to process acid foods in a water-bath canner varies from 5 to 85 minutes.

Caution: Always follow time, temperature, and pressure recommendations very carefully. In this book, they are given in each recipe.

Effect of Altitude. Altitude affects the temperature at which water boils. See Table 3 (p. 11) for the boiling point at various altitudes. This is important in canning food at home. Water boils at lower temperatures as altitude increases. The lower boiling temperatures are less effective for killing bacteria. If you live at altitudes of 2,000 feet or more, you may need to increase the processing time or canner pressure to compensate for the lower boiling temperatures. In Alabama, however, there are no altitudes above 1,000 feet, according to geological surveys from the University of Alabama.

Food Quality

The quality of canned food can be enhanced or reduced at any stage from harvest to storage, beginning with the variety you select. Some varieties of fruits and vegetables produce higher quality canned products than others that might be better when eaten fresh. Use the varieties you prefer that are better suited for canning. Your local Extension county agent can advise you on which canning varieties grow well in your area.

After harvesting, examine the food carefully to be sure it is fresh and wholesome. Discard old, diseased, or moldy food. Trim away any small diseased lesions or spots from food that is otherwise in good condition.

Can fruits and vegetables while they are still fresh. Most vegetables should be processed within 6 to 12 hours after being picked. However, apricots, nectarines, peaches, pears, and plums should be ripened 1 or more days between harvest and canning. Spread them out in a single layer on a clean, dry surface to ripen more evenly. If you must delay the canning of fresh produce, keep it in the refrigerator, if possible, or in a cool, dark place.

Fresh, home-slaughtered red meats and poultry should be chilled and canned without delay. Do not can meat or poultry from sickly or diseased animals. Place fish and seafood on ice at once after harvest; remove the internal organs immediately and can fish within 2 days.

To maintain good color and flavor during processing and storage:
• Use only high-quality foods which are at the proper stage of maturity and are free of diseases and bruises.
• Wash foods thoroughly and trim away small damaged areas.
• Use the hot-pack method (p. 30), especially for acid foods to be processed in a water-bath canner.
• After preparing foods for canning, protect them from too much exposure to light, heat, and air. Can them as soon as possible.
• When preparing enough for a full canner, keep peeled, halved, quartered, sliced, or diced apples, apricots, nectarines, peaches, and pears in an ascorbic-acid solution (see Pretreating, p. 39) to help keep them from turning dark. This will also help maintain the natural color of mushrooms and potatoes and prevent stem-end discoloration in cherries and grapes.
• Fill clean, hot jars with hot foods, leaving the amount of headspace specified in the recipes.
• Remove air bubbles from filled jars and wipe jar rims with a clean, damp cloth before adjusting lids.
• Use recommended self-sealing lids and screw bands or rings.
• Tighten screw bands securely but, if you are especially strong, not as tightly as possible. Read manufacturer's instructions.
• After canning, store sealed jars in a relatively cool, dark, dry place, preferably between 50° and 70°F.
• Can no more food than you will use within a year.

GENERAL STEPS IN CANNING

1. Preheat the canner by partially filling it with the appropriate amount of water (see Using the Water-Bath Canner, p. 32, or Using the Pressure Canner, pp. 32-34) and placing it over high heat. Heat another pot of water to boiling to add to the canner after adding jars.

2. Prepare standard canning jars (see Jars, pp. 21-25) and 2-piece lids (see Lids, pp. 25-26). Wash all jars in hot soapy water and rinse well. Then keep them hot until ready to fill. If using a water-bath canner to process foods for less than 10 minutes,

20 • Canning/General

sterilize the jars (see Sterilizing, p. 22) and keep them hot till ready to fill. You can keep jars hot by placing them in the water that is preheating in the canner. (See Equipment And Methods Not Recommended, p. 35.)

3. For foods to be canned in syrup or broth, prepare syrup (Table 9, p. 40), sauce, or broth in advance; keep it hot to avoid delay at canning time.

4. Select good quality, fresh food. Specific characteristics to look for in a type of food are described at the beginning of that food's recipe section. Work quickly from the time food is harvested or slaughtered to keep it fresh-tasting.

5. Sort pieces for size, color, and degree of maturity. Similar pieces in a jar make a better product.

6. Wash vegetables and fruits thoroughly until every trace of soil is removed. Trim away any small damaged or diseased spots from otherwise healthy food. Remove excess fat from meats and poultry. Remove head, tail, fins, and scales from fish. Prepare them according to the recipe you are using.

7. Fill hot jars (see Packing Methods, pp. 30-31). For hot pack, fill them quickly so that foods remain hot. Use hot liquid for both hot and raw packs that call for liquid. Use enough liquid, unless specified otherwise, to prevent a tight pack. Food should move slightly when jar is rotated. Cover with boiling water, syrup, or other recommended liquid, leaving the amount of headspace specified in your recipe (also, see Checking Headspace, pp. 31-32).

8. Work out air bubbles with a plastic knife or spatula.

9. Wipe jar rims—inside, outside, and top—with a clean, damp cloth or paper towel. Food or liquid on the rims can cause seals to fail.

10. Adjust lids and bands immediately on standard canning jars (see manufacturer's directions; also, see Lids, pp. 25-26).

11. Process food at once, according to directions in your recipe. (Also, see Using the Water-Bath Canner, p. 32, or Using the Pressure Canner, pp. 32-34).

12. Remove jars from a water-bath canner at the end of the processing time. Remove jars from a pressure canner as soon as the canner is depressurized. Let jars cool naturally, away from drafts, on a rack or cloth (see Cooling Jars, p. 34). Do not cover jars.

13. When jars are at room temperature, check to see if they are sealed (see Testing Jar Seals, p. 34). If not sealed, store in refrigerator or reprocess (see Reprocessing Food, pp. 34-35).

14. Do not remove bands on sealed jars until the next day. Then, wipe jars with a damp cloth and label with the name of the food and date.

15. Store canned foods in a cool (about 70°F), dark, dry place (see Storing Canned Foods, p. 35).

Caution: Boil all home-canned foods for 10 minutes before serving. Store unused portions in the refrigerator and use within 2 days for best quality.

MATERIALS AND EQUIPMENT

Needed:
- Standard canning jars
- Two-piece lids—self-sealing lid with metal rings
- Home canner—water-bath canner and/or pressure canner
- Colander
- Sieve
- Scales
- Measuring cup
- Plastic knife or spatula—for removing air bubbles
- Timer or clock
- Clean cloths—for wiping jar rims and cleaning up

Helpful:
- Jar lifter—helps remove hot jars from canner
- Jar funnel—helps pack small food items into jars
- Lid wand—helps remove treated lids dropped into hot water
- Quart measuring cup
- Clean towels
- Hot pads

Jars

Food may be canned in glass jars or metal containers. However, metal containers can be used only once, they require special sealing equipment, and they are much more costly than jars. Most Alabamians who enjoy home canning prefer to use jars.

Figure 1. Standard Canning Jars

Regular- and wide-mouth mason canning jars (see Figure 1, p. 21) are the best choice for home canning. They are available in ½-pint, 1-pint, 1½-pint, 1-quart, and ½-gallon sizes. Half-gallon jars are best used for canning very acid juices. Decorator jelly jars are available in 6- and 12-ounce sizes. The regular-mouth jar opening is about 2⅜ inches in diameter. Wide-mouth jars have openings of about 3 inches, making them easier to clean, fill, and empty.

With careful use and handling, mason jars may be reused many times for up to 13 years. When these jars are used properly with self-sealing lids, jar seals and vacuums are excellent and jar breakage is rare.

Most commercial pint- and quart-size mayonnaise or salad dressing jars may be used with new, 2-piece, self-sealing lids for canning acid foods in the water-bath canner. However, you can expect more seal failures and jar breakage. These jars have a narrower sealing surface and are not tempered (made tough) as well as mason jars. Also, they are often weakened by repeated contact with metal spoons or knives that had previously been used to remove servings of the mayonnaise or salad dressing. Seemingly insignificant scratches in the glass may cause cracking and breakage while jars are processing in a canner.

Caution: Mayonnaise-type jars should not be used in a pressure canner because of excessive jar breakage. Other commercial jars, such as peanut butter, pickles, etc., are not recommended for use in canning any food at home because they cannot be sealed with the 2-piece self-sealing lids.

Cleaning. Wash empty jars in hot water with detergent and rinse well by hand, or wash in a dishwasher. Be certain to rinse carefully because detergent residue can cause unnatural flavors and colors. Scale or hard-water film on jars is easily removed by soaking jars several hours in a solution of 1 cup vinegar (5% acidity) per gallon of water. These washing methods do not sterilize jars.

Sterilizing. Sterilize jars for all jams, jellies, and pickled products that will be processed for less than 10 minutes. To sterilize empty jars, put them right side up on a rack in a water-bath canner or large pot. Fill the jars and canner with hot (not boiling) water to 1 inch above the tops of the jars. Heat to boiling and boil 10 minutes.

Jars used for vegetables, meats, and fruits that will be processed in a pressure canner do not need to be sterilized before filling. It is also unnecessary to sterilize jars for fruits, tomatoes, and pickled or fermented foods that will be processed 10 minutes or longer in a water-bath canner. The jars and foods will be sterilized during the processing. However, you may want to boil them anyway to test the jars for imperfections that may lead to breakage (see Examining and Testing, below).

Examining and Testing. Examine tops of new as well as old jars to see that there are no nicks, cracks, sharp edges, or dips. Run your finger around the sealing surface of each jar. Also, examine the body and bottom of each jar for any other imperfections. If there is a crack or dip in a new jar, return it to the manufacturer for a refund. The slightest imperfection can cause a sealing failure.

Test new jars for imperfections that might lead to jar breakage the same way you sterilize them. Place jars right side up on a rack in a water-bath canner or large pot. Fill jars and canner with hot (not boiling) water to 1 inch above the tops of the jars. Heat to boiling and boil 10 minutes.

Jar Breakage. Jars occasionally break during the canning process. Understanding the reasons for breakage can help you prevent it. There are three major types of jar breakage as well as some other reasons.

Thermal Shock Breakage. This type is characterized by a crack running around the base or lower part of the jar and sometimes extending up the side. It can be caused by any of the following:

- Sitting hot jars to cool on a cold surface or near a cold draft.
- Accidentally spattering hot jars with cold water.
- Not using a rack in the water-bath or pressure canner.
- Filling hot jars of food with water or syrup that is not boiling.
- Filling hot jars with cold food and placing them into boiling water.
- Using soap-impregnated steel wool pads to clean jars or using a sharp knife to remove air bubbles and making fine scratches in the glass.

Pressure Breakage. A vertical crack on the side which divides and forks into fissures indicates pressure breakage. It can be caused by any of the following:

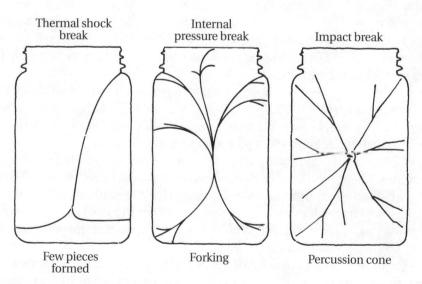

Figure 2. Three Types of Jar Breakage

- Using the oven to process home-canned foods.
- Using too shallow a headspace and, therefore, not leaving enough space to permit expansion of the contents during heating (or freezing).
- Not keeping the heat steady under the pressure canner.
- Touching the petcock or weighted control after processing but before all pressure has released from the canner.
- Pouring cold water over the canner and, thereby, not allowing it to cool down naturally.

Impact Breakage. This is a break originating at the point of impact with fissures radiating from that point. It can be caused by either of the following:
- Using jars that have received trauma from rough handling while being transported or used at home, such as dropping, hitting, or bumping.
- Hitting the jar with a sharp metal knife to remove air bubbles.

Other Types. Breakage can be caused by any of the following:
- Using jars that are too old; canning jars have a life expectancy of about 13 years, if cared for properly.
- Using jars with imperfections. Imperfect jars will break the first time they are used. It is a good idea to test all new jars before using (see Examining and Testing, pp. 22-23).
- Using commercial jars (such as mayonnaise jars) instead of mason jars in a pressure canner. Commercial jars sometimes break in a water-bath canner as well.

Advantages of Using Standard Canning Jars. To be sure of good quality home-canned foods, use standard mason canning jars. You can use commercial mayonnaise jars, but you are taking a risk. Buying standard canning jars is a good investment because, if cared for properly, they can be reused for up to 13 years. The following are more reasons why standard canning jars are best for home canning:
- Standard canning jars have a deeper finish than regular jars which provides a tighter seal.
- Manufacturers of standard canning jars will refund your money if their jars have imperfections when you buy them.
- All brands of mason jars have been standardized to meet the same specifications. Therefore, they are all designed for the 2-piece self-sealing lids and to withstand the temperature and pressure of the canners. Also, canning recipes are written for the jar sizes and types. Commercial jars, such as mayonnaise jars, are made to the specifications of the product (mayonnaise) that was originally packed in it.
- Commercial jars have a wide range of difference in the diameter of the mouth and the width of the sealing surface. Lids aren't as likely to fit and seal well.

- The glass lugs on commercial jars are designed to be used with different types of screw caps or bands.
- High-speed conveyor-belt filling lines traumatize the commercial jars during their filling process and can cause the jars to break during home processing.

Lids

The recommended 2-piece, self-sealing canning lid consists of a flat metal lid held in place by a metal screw band. The flat lid is crimped around its bottom edge to form a groove that is filled with a gasket compound. When jars are processed in the canner, the compound softens and flows slightly to cover the jar's sealing surface while still allowing air to escape from the jar. As the jar cools after processing, the contents in the jar contract, pulling the self-sealing lid firmly against the jar to form a high vacuum. The gasket compound hardens and forms an airtight seal.

Unused lids will keep for about 5 years from the date of manufacture. If the gasket compound is older, lids may fail to seal on jars. For best results, buy only the number of lids you will use in one year.

To ensure a good seal, carefully follow the manufacturer's directions in preparing lids for use. Examine all metal lids carefully. Do not use old, dented, or deformed lids or lids with gaps or other defects in the gasket compound.

Don't boil lids. **Pour boiling water over lids** to scald them to remove surface bacteria. Molds, yeasts, and vegetative cells of bacteria are killed at 170°F.

After filling jars, clean each jar's sealing surface with a clean, damp cloth and place the lid, gasket down, onto the cleaned sealing surface. Then, fit the metal screw band over the flat lid.

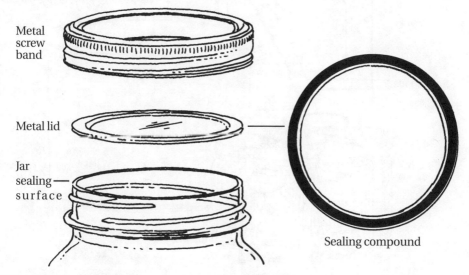

Figure 3. Two-Piece Lid

26 • Canning/General

Follow the manufacturer's guidelines enclosed with or on the box for tightening the screw bands properly. **If bands are too loose,** liquid may escape from jars during processing and seals may fail. **If bands are too tight,** air cannot vent during processing, and food will discolor during storage. Overtightening may also cause lids to buckle and jars to break, especially with raw-packed food in a pressure canner.

Caution: Do not retighten lids after processing jars.

Screw bands are not needed on jars of food in storage. They can be removed easily after jars are completely cooled, at least 12 hours after processing. When removed, washed, dried, and stored in a dry area, screw bands may be used many times. If screw bands are left on stored jars, they may become rusty and difficult to remove, and they may not work properly again.

Canners

Canning equipment for processing food with heat includes two main types: water-bath canners and pressure canners. Most are designed to hold seven quart jars or eight to nine pint jars. There are some large (deeper) pressure canners that hold eighteen pint jars in two layers, but they still hold only seven quart jars.

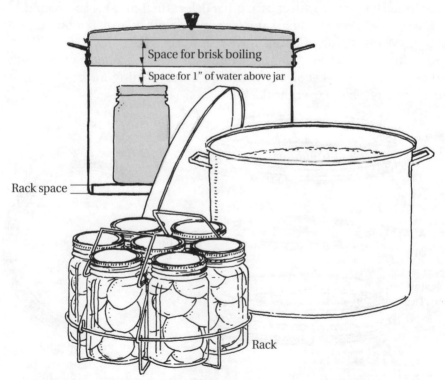

Figure 4a. Water-Bath Canner

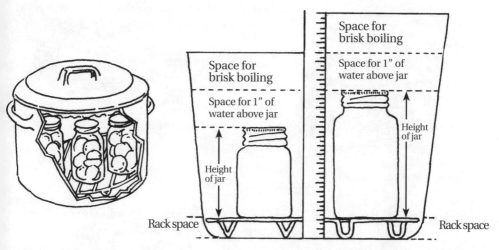

Figure 4b. Water-Bath Canner

Small pressure canners hold four quart jars. They are operated basically the same as the standard-size canners and must be vented using the typical venting procedures.

Pressure saucepans with smaller volume capacities are not recommended for canning jars larger than ½ pint.

Pressure canners must be used to process low-acid foods to prevent the risk of botulism. Pressure canners may also be used for processing acid foods; however, water-bath canners are usually recommended because they are faster. A pressure canner requires from 55 to 100 minutes to process most acid foods; a water-bath canner requires from 25 to 75 minutes. For example, a water-bath canner loaded with filled jars must heat about 20 to 30 minutes before its water begins to boil. Then, it must process the food for 5 to 45 minutes, depending on the food. In contrast, a loaded pressure canner must heat about 12 to 15 minutes before it begins

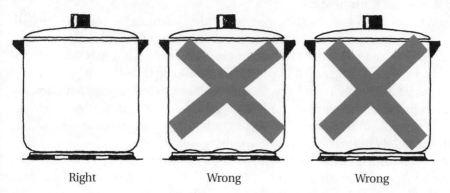

Figure 5. Flat- and Ridged-Bottom Canners on Electric Burners

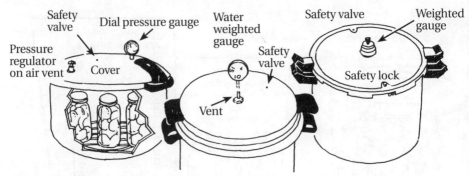

Figure 6. Pressure Canners

to vent; then, it takes another 10 minutes to vent the canner, 5 minutes to pressurize the canner, 8 to 10 minutes to process the acid food, and finally, 20 to 60 minutes to cool the canner before removing jars.

Water-Bath Canners. Standard water-bath canners are made of aluminum or porcelain-covered steel. They have removable perforated racks, usually with handles, and fitted lids. This canner is designed to be deep enough so that at least 1 inch of briskly boiling water will be over the tops of jars without boiling over during processing.

Some water-bath canners have ridged bottoms and some have flat bottoms. **A flat-bottom canner must be used on an electric range.** Either a flat or ridged bottom can be used on a gas burner. To ensure uniform processing of all jars with an electric range, the canner should be no more than 4 inches wider in diameter than the burner on which it is heated.

Most any large pot with a tight-fitting lid can be used for boiling (or simmering) water-bath processing, but it must be deep enough to hold a rack, the jars, and still have room to cover the jars by at least 1 inch of briskly boiling water without boiling over. If you do not have a rack, place the jars on heavy dish cloths to keep them from touching the bottom of the pot.

Canner Care. Wash and dry the canner after each use. To remove the dark water stain that may develop inside an aluminum canner, fill it to above the stained area with a mixture of 1 tablespoon cream of tartar for each quart of water. Heat to boiling and boil until stain disappears. Then, wash again with hot soapy water, rinse, and dry. Store it with crumpled newspaper or paper towels in the bottom and around the rack to absorb any moisture or odors.

Pressure Canners. These canners have been extensively redesigned since 1970. Models made before 1970 are heavy-walled pots with clamp-on or turn-on lids. They are fitted with a dial gauge, a vent port in the form of a petcock or counterweight, and a safety fuse. If properly cared for and annually checked, these older models can still perform well.

Pressure canners made since 1970 are lighter weight pots with thinner walls; most have turn-on lids. They have a jar rack, gasket, dial or weighted gauge, an automatic vent/cover lock, a vent port (steam vent) to be closed with a counterweight or weighted gauge, and a safety fuse.

Pressure does not destroy microorganisms, but it creates the high temperature that does destroy them when applied for an adequate period of time. The success of destroying all microorganisms capable of growing in low-acid canned foods depends on reaching a temperature of 240° to 250°F and holding it for the time required for the specific food being canned.

Two serious errors that may occur in pressure canners do so when:

1. The canner is operated at an altitude that is 2,000 feet or more above sea level (see Effect of Altitude, p. 18). To correct this error, food must be processed for a longer time or at increased pressure. This is not a problem in Alabama because no area of the state is as much as 1,000 feet above sea level.

2. Too much air is trapped inside the jars or the canner. This air lowers the temperature and results in underprocessing. Higher volumes of air become trapped in a canner when using the raw-pack method and dial-gauge canners. Dial-gauge canners do not vent air during processing.

Removing Air. All pressure canners must be **exhausted or vented** to remove air. Air interferes with the passage of heat; heat passes through steam more readily than through air. To vent a canner, leave the weight off the vent port or manually open the petcock on some older models. Heating the filled canner at its highest setting with its lid locked into place generates steam that escapes through the vent port or petcock. When steam first escapes, set a timer for 10 minutes. Adjust heat so air travels through the vent tube at a slow to moderate rate. After venting 10 minutes, close the petcock or place the counterweight or weighted gauge over the vent port to pressurize the canner.

When **weighted-gauge** models are pressurized, they exhaust tiny amounts of air and steam each time their gauge rocks or jiggles during processing. They control pressure precisely at 5, 10, or 15 pounds of pressure at sea level. Most recipes in this book require 10 pounds of pressure. The sound of the weight rocking or jiggling indicates that the canner is maintaining the recommended pressure. The single disadvantage of weighted-gauge canners is that they cannot be corrected precisely for higher altitudes.

Tests on **dial-gauge canners** have shown they register a slightly different pressure from weighted-gauge canners. For that reason, dial gauges should be operated at 11 pounds pressure instead of 10 at altitudes up to 2,000 feet. Check dial gauges for accuracy before each year's use and replace them if necessary. Gauges may be checked at most Extension System county offices.

Lid **safety valves** are thin metal inserts or rubber plugs designed to relieve excessive pressure from the canner. Do not pick at or scratch valves while cleaning lids.

Use only canners that have the **Underwriter's Laboratory (UL) approval to ensure their safety.**

Canner Care. Handle canner lid gaskets carefully and clean them according to the manufacturer's directions. Nicked or dried gaskets will allow steam leaks during pressurization of canners. Keep gaskets clean between uses. Gaskets on older model canners may require a light coat of vegetable oil once each year. Gaskets on later model canners are prelubricated and do not benefit from oiling. Check your canner's use and care booklet for specific instructions.

For safe operation, the vent, safety valve, and edges of the lid and canner must be clean at all times. To clean the vent, draw a string or narrow strip of cloth through the opening.

After use, wash the canner with hot, soapy water, being careful not to immerse the dial gauge if your canner has one. Rinse and dry well. To remove the dark water stain that may develop inside an aluminum canner, fill it to above the stained area with a mixture of 1 tablespoon cream of tartar for each quart of water. Heat to boiling and boil until stain disappears. If the stain is stubborn, use more cream of tartar. Then, wash again with hot soapy water, rinse, and dry well. Store the canner with crumpled newspapers or paper towels in the bottom and around the rack to help absorb any moisture and odors. Place the lid upside down on the canner. Never put the lid on the canner and seal it for storage.

METHODS AND PROCEDURES

Packing Methods

Many fresh foods contain from 10 to more than 30 percent air. Removing air from food before jars are sealed is important for good quality. Hot-packing is the best way to remove air and is the preferred way for packing food to be processed in a water-bath canner.

Hot-Pack. This method requires heating freshly prepared food to boiling, simmering it 2 to 5 minutes, and promptly filling jars loosely with the boiled food. It helps remove air from food tissues, shrinks food, helps keep the food from floating in the jars, increases the vacuum in sealed jars, and improves shelf life of the canned food. Shrinking food allows you to get more into each jar. At first, the color of hot-packed foods may appear no better than that of raw-packed foods, but within a short storage period, both color and flavor of hot-packed foods will be better.

Raw-Pack. Also called cold-pack, this method requires filling jars tightly with freshly prepared, but unheated, food and adding hot liquid. Oftentimes, raw-packed food, especially fruit that is processed in a water-bath canner,

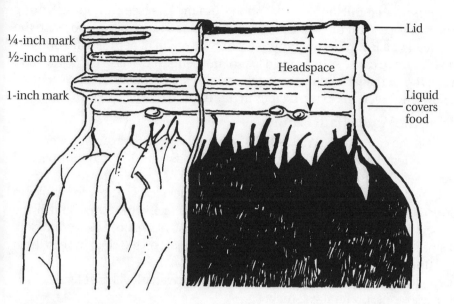

Figure 7. Headspace

will float in the jar after processing because air was trapped in the raw food. The trapped air in and around the food may cause discoloration within 2 to 3 months of storage. Raw-packing is more suitable for vegetables to be processed in a pressure canner because the pressure will create a vacuum pack.

Caution: Whether food is to be hot-packed or raw-packed, the juice, syrup, broth, or water that is added to the food should be boiling when adding it to the jars.

Checking Headspace

The unfilled space at the top of a jar, between the food and the lid, is called headspace. Directions for canning specify leaving a ¼-inch headspace for jellies and other jellied fruit spreads, ½-inch for fruits and tomatoes that will be processed in a water-bath canner, and from 1 to 1¼ inches for low-acid foods that will be processed in a pressure canner.

Hold the jar at eye level to check for accurate headspace. Looking over the top of a jar will not give you an accurate measurement. On standard canning jars (see Figure 7, above):

- The ¼-inch mark is just below the first glass lug.
- The ½-inch mark is just below the middle glass lug.
- The 1-inch mark is the line below the bottom glass transfer bead.

Headspace is needed for foods to expand while they are being processed and for forming a vacuum in cooled jars. The extent of expansion depends on the amount of air in the food and the processing temperature. Air expands greatly when heated to high temperatures. Food expands less than air when heated.

If jars are too full, liquid will siphon from the jars (see Siphoning During Processing, p. 36). **If the headspace is too deep,** too much air will be left in the jar, creating a favorable environment for the growth of aerobic microorganisms. This can cause spoilage and loss of nutrients. It can also cause the food at the top of the jar to darken and the seal to fail because of a low vacuum. Follow the directions in the recipes in this book for correct headspace.

Using the Water-Bath Canner

1. Fill the canner (see Water-Bath Canners, p. 28) halfway with water.

2. Preheat water to 140°F for raw-packed foods or to 180°F for hot-packed foods (see Packing Methods, pp. 30-31).

3. Load filled jars, fitted with lids, onto the canner rack and use the handles to lower the rack into the canner. Or, place the canner rack in the canner and fill the canner, one jar at a time, with a jar lifter. If you have no rack, place a heavy dish cloth or small towel in the bottom of the pot and sit the jars on the cloth to keep them from touching the canner bottom.

4. Add more boiling water, if needed, so the water level is at least 1 inch above jar tops, preferably 2 inches.

5. Turn heat to its highest position until water boils vigorously.

6. When water boils, set a timer for the minutes required for processing the food, specified in your recipe.

7. Cover with the canner lid and lower the heat setting to maintain a gentle boil (212°F) throughout the processing time.

8. Check the water level periodically and add more boiling water, if needed, to keep water at least 1 inch above the jars.

9. When filled jars have been boiled for the recommended time, turn off the heat and remove the canner lid carefully. Tilt the lid away from you as you lift it so that the steam will not burn your face or arms as it escapes from the canner (see Figure 8).

10. Using a jar lifter, remove the jars and place them, at least 1 inch apart, on a towel or cooling rack (see Cooling Jars, p. 34).

Using the Pressure Canner

1. Put 2 to 3 inches of hot water in canner (see Pressure Canners, pp. 28-30). Place filled jars on rack, using a jar lifter. Fasten canner lid securely.

2. Leave the weight off the vent port or open the petcock. Then, heat to the highest setting until steam flows from the petcock or vent port. Maintain the high heat setting to exhaust steam for 10 minutes.

Figure 8. Lifting Lid to Avoid Steam Burn

3. After exhausting steam, place weight on vent port or close the petcock. The canner will pressurize during the next 3 to 5 minutes.

4. When the pressure reading on the dial gauge indicates that the recommended pressure has been reached or when the weighted gauge begins to jiggle or rock, set a timer for the minutes required for processing.

5. Regulate the heat under the canner to maintain a steady pressure at or slightly above the correct gauge pressure. Quick or wide variations of pressure during processing may cause unnecessary liquid losses from jars (see Siphoning During Processing, p. 36). Weighted gauges on Mirro canners should jiggle about 2 or 3 times per minute. On Presto canners, they should rock slowly throughout the process.

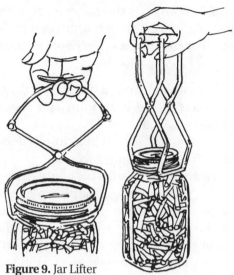

Figure 9. Jar Lifter

Note: Trade or brand names used in this book are for information only. The Alabama Cooperative Extension System does not guarantee nor warrant the reliability of any product mentioned.

6. When the timed process is completed, turn off the heat and remove the canner from the heating unit, if possible, to let the canner depressurize.

Caution: Do not force-cool the canner. Cooling the canner with cold running water or opening the vent port before the canner is fully depressurized will cause loss of liquid from jars and seal failures, resulting in food spoilage. Force-cooling may also cause jars to break or may warp the canner lid of older model canners, causing steam leaks. Depressurization of older models should be timed: standard-size, heavy-walled canners require about 30 minutes when loaded with pints and 45 minutes with quarts. Newer, thin-walled canners cool more rapidly and are equipped with vent locks. These canners are depressurized when their vent lock piston drops to a normal position.

7. After the canner is depressurized, remove the weight from the vent port or open the petcock. Wait **2 minutes** before unfastening the lid. Then, remove it carefully, tilting the lid away from you as you lift it so that the steam will not burn your face or arms as it escapes from the canner (see Figure 8).

Caution: It is important that you not delay removing the lid beyond the time specified above. If you wait, a vacuum may develop in the canner, and it may be impossible to remove the lid.

8. Remove jars as soon as you remove the lid to prevent the growth of heat-loving bacteria. Remove the jars with a jar lifter and place them at least 1 inch apart on a towel or cooling rack (see Cooling Jars, below).

Cooling Jars

Remove jars from a water-bath canner as soon as the processing time is complete. Remove jars from a pressure canner as soon as the pressure is out of the canner and the lid is removed (see Using the Pressure Canner, steps 7-8, pp. 33-34). The heat in a closed canner is a favorable environment for the growth of heat-loving bacteria.

When you remove the jars from the canner, do not tighten the jar lids. Tightening hot lids may cut through the gasket compound and cause seals to fail. Cool the jars at room temperature for 12 to 24 hours. Jars may be cooled on racks or towels to prevent heat damage to counters. Place them about 1 inch apart to allow good air circulation for cooling. But do not place them in a draft or near an open window.

If a jar has lost excessive liquid during processing, do not open it to add more liquid. As long as the jar is properly sealed (see Testing Jar Seals, below), the food is safe to use.

Testing Jar Seals

After cooling jars for 12 to 24 hours, remove the screw bands and test seals using one of the following methods:

1. Press the middle of the lid with your finger or thumb. If the lid springs up when you release your finger, the lid is not sealed.

2. Tap the lid with the bottom of a teaspoon. If it makes a dull sound, the lid is not sealed. However, if any food is touching the underside of the lid, it will cause a dull sound. If the jar is sealed correctly, it will make a ringing, high-pitched sound.

3. Hold the jar at eye level and look across the lid. The lid should be concave (curved down slightly in the center). If the center is either flat or bulging, it may not be sealed.

Reprocessing Food

If a lid fails to seal on a jar, remove the lid and discard it. Check the jar's sealing surface for tiny nicks. Check the sides and body of the jar for hairline cracks. If the jar is nicked or cracked, throw away the food and the jar. Bits of glass could possibly be in the food.

If the jar is not nicked or cracked, wipe the jar mouth with a clean, damp cloth or paper towel. Add a new, properly prepared lid and band, and process the food again within 24 hours, using the same instructions, temperature, and time.

A single unsealed jar could be safely stored in the refrigerator and the food eaten within several days. Also, unsealed jars may be stored in the freezer. Allow a 1½-inch headspace in jars stored in the freezer for expansion in freezing.

Storing Canned Foods

If lids are tightly vacuum sealed (see Testing Jar Seals, p. 34), remove the screw bands and carefully wash the filled jars and lids to remove any food residue. Then, rinse and dry the jars. Label and date the jars and store them in a clean, cool, dark, dry place. Do not store jars above 95°F or near hot pipes, a range, a furnace, in an uninsulated attic, or in direct sunlight. Under such conditions, food will lose quality in a few weeks or months and may spoil. Dampness may corrode metal lids, break seals, and allow contamination and spoilage (see Spoilage During Storage, p. 36).

Accidental freezing of canned foods will not cause spoilage unless jars become unsealed and contaminated. However, freezing and thawing may soften food. If jars must be stored where they may freeze, wrap them in newspapers, place them in heavy cartons, and cover with more newspapers and blankets.

CANNING PROBLEMS

Equipment and Methods Not Recommended

Open-kettle canning (with no lid on the canner) and the processing of freshly filled jars in **conventional ovens, microwave ovens,** and **dishwashers** are not recommended. These practices do not prevent all risks of spoilage.

Steam canners are not recommended because processing times for use with current models have not been adequately researched. Because steam canners do not heat foods in the same manner as boiling water-bath canners, their use with water-bath canner processing times may result in spoilage.

Pressure in excess of 15 PSI is not recommended when using new pressure-canning equipment.

So-called **canning powders** are useless as preservatives and do not replace the need for proper heat processing.

Jars with wire bails and glass caps make attractive antiques or storage containers for dry food ingredients but are not recommended for use in canning. **One-piece zinc porcelain-lined caps** are also no longer recommended. Both glass and zinc caps use flat rubber rings for sealing jars which too often fail to seal properly.

Siphoning During Processing

During the processing period, air is drawn from the jars in a pressure canner. Sometimes, liquid is also drawn from the jars. This is called siphoning.

When this happens, food particles can lodge in the lid's gasket compound, causing the seals to fail.

If grease gets into the gasket compound, the compound will break down. When this happens, the lid will stick to the jar and cannot vent up and down. Pressure builds up inside the jar and steam cuts a channel through the compound as it exhausts from the jar. This leaves a light impression on the compound, and the lid may buckle.

The following conditions can cause siphoning:
- Jars are too full; there is not enough headspace.
- Liquid was added to uncooked meat.
- Canner is exhausted too rapidly.
- Temperature is lowered or raised during processing time to maintain 10 pounds pressure.
- Pressure is too high for processing; the control regulator shakes more than three or four times per minute.
- The control regulator (petcock) is touched before the pressure leaves the canner naturally.
- Cold water is poured over the canner before pressure is out.
- Canner has a steam leak; it needs a new gasket.
- Dial gauge is faulty; canner was not checked correctly prior to canning season.

Spoilage During Storage

Do not taste stored, home-canned food from a jar with an unsealed lid or food that shows signs of spoilage. Growth of some types of spoilage bacteria and yeast produces gas which pressurizes the food, swells the lids, and breaks jar seals.

Check each jar of home-canned food before using it:

1. As each stored jar is selected for use, examine its lid for tightness and vacuum. Lids with concave centers have good seals.

2. While holding the jar upright at eye level, rotate it and examine its outside surface for streaks of dried food originating at the top of the jar. Look at the contents for rising air bubbles and unnatural color.

3. While opening the jar, smell for unnatural odors and look for spurts of liquid and cotton-like mold growth (white, blue, black, or green) on the surface of the food and underside the lid.

Caution: Signs of spoilage on low-acid foods, including tomatoes, don't always look the same. Or, you may see little or no sign at all. Therefore, any suspected container of such foods, should be treated as though the *botulinum* toxin had developed (see Disposal of Spoiled Canned Foods, p. 37).

Disposal of Spoiled Canned Foods

Spoiled canned foods must be handled with care in one of the following ways:

1. If the suspected jars are still sealed, place them in a heavy garbage bag. Close the bag and place it in a regular trash container or bury it in a nearby landfill.

2. If the suspected jars are unsealed, open, or leaking, detoxify them before disposal (see Detoxification Process, below).

Detoxification Process

1. Carefully remove the lids and place the suspected jars, their contents, and lids in an 8-quart or larger pot, saucepan, or water-bath canner. Lay the jars on their side. Wash your hands thoroughly with soap and hot water after handling the jars. Carefully add water to the pot; avoid splashing the water. The water should completely cover the containers by at least 1 inch.

2. Place a lid on the pot and heat the water to boiling. Boil 30 minutes to ensure detoxifying the food, the jars, and lids. Cool and **discard** the jars, their lids, and the food in the trash or bury them in soil to be sure no animal or human can get the food.

3. Thoroughly scrub all counters, containers, and equipment, including the jar opener, your clothing, and hands that may have contacted the food or jars. Discard any sponges or wash cloths that may have been used in the clean-up. Place them in a plastic bag and discard them in the trash.

Tips to Prevent Problems

The essential elements in processing canned foods:
- Select good foods.
- Prepare them correctly.
- Use recommended equipment.
- Follow directions carefully.
- Process at the correct temperature for the correct time.
- Store canned foods properly.
- Don't take shortcuts!

Common Causes of Problems

Spoilage. DO NOT TASTE ANY HOME-CANNED FOOD THAT HAS SIGNS OF SPOILAGE! Home-canned foods may spoil if
- the food is packed too tightly in the jar
- the food is not processed long enough
- the pressure in a pressure canner is incorrect for the food
- the pressure gauge is not accurate
- the pressure canner is not exhausted adequately
- the water is too low in the canner
- the jar lid does not seal

Dark or Discolored Foods. Color changes in foods can occur when
- the foods are underripe or overripe
- iron, zinc, or copper utensils are used in preparing foods
- the liquid does not cover the food in the jar
- air bubbles are not removed
- the food is not processed long enough
- the food is processed too long
- the temperature is too low during processing
- the water is too low in the canner
- the jar is not properly sealed
- the food is stored in too much light
- spoilage occurs

Floating Foods. Foods, especially fruits, sometimes float in their juice when
- the fruit is overripe
- the food is not adequately cooked before filling the jar
- the food is packed too loosely in the jar
- the syrup is too heavy (too much sugar) for the fruit
- the food is processed too long
- the food is processed at too high a temperature

Loss of Liquid From Jar. Liquid sometimes disappears or siphons from the jar because
- the food is packed too tightly
- air bubbles are not removed
- not enough headspace is left in the jar
- starchy foods absorb all the liquid
- the water boils too fast in a water-bath canner
- the water is too low in the canner
- the pressure fluctuates during processing
- the pressure is lowered too quickly in a pressure canner

Cloudy Liquid or Sediment. Cloudiness and/or sediment sometimes appears if
- the vegetable is immature
- table or iodized salt is used
- the fruit is overripe
- there are minerals in the water
- spoilage occurs

Note: A natural, harmless yellow sediment appears in some canned fruits and vegetables.

CANNING FRUITS

Selecting

Use only clean, ripe, good quality fruits for canning. Only grapes that are to be canned whole should be underripe. For most fruits, choose those that are firm and ideal for eating fresh. Apples should be juicy and crisp. Fruits and berries that will be canned as juice should be well colored. Discard all fruits that are insect-damaged, diseased, discolored, or severely bruised. Small spots of damage or disease on otherwise healthy fruits can be trimmed away.

Pretreating

Most light-colored fruits, such as apples, pears, and peaches, and some vegetables, such as potatoes and mushrooms, will turn dark or become discolored when they are peeled or cut into pieces and their flesh is exposed to air. This can be prevented by pretreating them. The best method of pretreating fruits that will be canned is by dipping peeled or cut fruits into an ascorbic acid solution. You can get ascorbic acid in several forms.

Pure Powdered Form. This is available in supermarkets during the canning season with other canning supplies. One level teaspoon of pure powdered ascorbic acid weighs about 3 grams. Use a solution of 1 teaspoon powder per gallon of water.

Vitamin C Tablets. These are economical and available year-round in many stores. Buy 500-milligram tablets. Crush and dissolve six tablets per gallon of water.

Commercially Prepared Mixes. Ascorbic and citric acid is available in supermarkets during canning season. Sometimes citric acid powder is sold in supermarkets, but it is less effective in controlling discoloration. If you choose to use these products, follow the manufacturer's directions.

Canning

Fruits may be canned in either a boiling water-bath or a pressure canner. Unlike vegetables, most fruits have enough natural acid to be quickly and safely canned in a water-bath canner. The total time needed for the canning process is much longer in a pressure canner because of the extra time needed to heat up, exhaust, pressurize, and cool down.

40 • Canning/Fruits

In Syrup. Adding syrup to fruit when canning helps the fruit to retain its flavor, color, and shape. However, it does not prevent spoilage, as some people have thought. Table 9 (below) includes guidelines for preparing and using syrups, ranging from very light to very heavy and increasing in sugar content by about 10 percent each.

The very light syrup approximates the natural sugar content of many fruits. It contains fewer calories and makes a tasty product of many fruits that have usually been packed in heavy syrup. Try a small amount the first time to see if you like it.

The amount of syrup needed to fill standard canning jars depends on the type of pack. The approximate amount of syrup required is ⅔ cup for pint jars and 1 to 1⅓ cups for quart jars. Quantities of water and sugar given in Table 9 will make enough syrup for a canner load of 9 pints, 4 quarts, or 7 quarts. Directions for making the syrup are at the end of the table.

Table 9. Syrups for Canning

Syrup Type	Approx. % Sugar	Number of Cups to Use*				Comments and Fruits Commonly Packed in This Syrup
		For 9 Pints or 4 Quarts		For 7 Quarts		
		Water	Sugar	Water	Sugar	
Very Light	10	6½	¾	10½	1¼	Approximates natural sugar level in most fruits.
Light	20	5¾	1½	9	2¼	For very sweet fruits.
Medium	30	5¼	2¼	8¼	3¾	For sweet apple varieties, sweet cherries, berries, grapes, sweet potatoes.
Heavy	40	5	3¼	7¾	5¼	Tart apples, apricots, sour cherries, pears, peaches, plums, gooseberries, nectarines.
Very Heavy	50	4¼	4¼	6½	6¾	For very sour fruits.

*The approximate amounts of syrup required per jar are ⅔ cup per pint and 1 to 1⅓ cups per quart. The exact amount depends on the type of pack.
To Make Syrup: Combine water and sugar and heat to a boil, stirring to dissolve sugar. Use syrup as directed in recipe you are using.
Note: Light corn syrup or mild-flavored honey may be used to replace up to half the table sugar called for in syrups.

Table 10. Pounds of Fresh Fruits Needed for Standard Canner Loads

Type and Form of Fruit	Pounds of Whole Fruit in 1 Bushel	Pounds of Whole Fruit Needed to Can:		
		7 quarts	9 pints	1 quart
Apples, sliced	50	19	12¼	2¾
Applesauce	50	21	13½	3
Apricots, halved, sliced	50	16	10	2¼
Berries, except Strawberries and Cranberries	36+	12	8	1¾
Cherries, whole	56	17½	11	2½
Cranberries	100	7	4½	1
Figs, whole		16	11	2½
Grapefruit, sections	40++	15	13	2
Grapes, whole	48	14	9	2
Grape Juice	48	24½	16	3½
Nectarines, halved, sliced	48	17½	11	2½
Oranges, sections		15	13	2
Peaches, halved, sliced	48	17½	11	2½
Pears, halved, quartered	50	17½	11	2½
Pineapple, sliced, cubed	70+	21	13	2½
Plums, whole, halved	56	14	9	
Rhubarb, stewed	28+++	10½	7	1½
Strawberries	36+			
Tomatoes, whole, halved	53	21	13	3
Tomatoes, crushed	53	22	14	2¾
Tomato Juice	53	23	14	3¼
Tomato Sauce,				
thin	53	35	21	5
thick	53	46	28	6½

+Pounds of whole berries in a 24-quart crate rather than a bushel.
++Pounds of whole fruit in a bag rather than a bushel.
+++Pounds of whole fruit or berries in a lug rather than a bushel.

In Juice. Unsweetened commercial apple juice, pineapple juice, and white grape juice make good liquids for packing fruits. Dilute with water, if desired. You may also use the juice extracted from some fruits being canned.

Without Sugar. If you want to can fruits without sugar, it is very important to select fully ripe but firm fruits of the best quality. Prepare them as described in the recipes and be sure to use the hot-pack method described for each specific fruit. Also, use water or regular unsweetened fruit juice instead of a sugar syrup. Juice made from the fruit being canned is best to use. Blends of unsweetened apple, pineapple, and white grape juice are also good for most fruits. Pack jars and process the same as described in the recipes. Use sugar substitutes, if desired, only when serving the canned fruits, not when canning. Canning with sugar substitutes sometimes cause the fruits to become bitter during storage.

For Baby Food. Any fruit prepared for baby food in chunk style or pureed, with or without sugar, may be canned using the recipes for the specific fruits. Use the hot-pack method and pack in half-pint, preferably, or pint jars. Process for 20 minutes in a boiling water-bath canner.

Caution: Heat all home-canned foods to boiling and boil 10 minutes before serving. Store unused portions in the refrigerator and use within 2 days for best quality.

FRUIT RECIPES

APPLE JUICE

Use a blend of apple varieties to make good quality apple juice. See Table 10 (p. 41) for amount. For best results, begin the following procedure within 24 hours after juice has been pressed from the apples. See General Steps In Canning (pp. 19-20). Refrigerate juice for 24 to 48 hours. Without mixing, carefully pour off clear juice and discard sediment. Strain clear juice through a paper coffee filter or double layers of damp cheesecloth into a large, thick-bottom saucepan.

Hot Pack: Heat juice quickly, stirring occasionally, until it begins to boil. Fill hot jars immediately with juice, leaving a ¼-inch headspace. Wipe jar rims and adjust lids.

Process in a boiling water-bath canner.
Pints and Quarts, 5 minutes
Half-gallons, 10 minutes

APPLE SLICES

Use apples that are juicy and crisp. Combine both sweet and tart varieties, if possible. See Table 10 (p. 41) for amount. See General Steps In Canning (pp. 19-20). Wash apples and remove stems. To help prevent darkening, prepare only one apple at a time. Peel, core, and slice each apple into a bowl containing an ascorbic acid solution (see Pretreating, p. 39). Apples may be canned in syrup, apple or pineapple juice, white grape juice, or water. If canning in syrup, prepare a very light, light, or medium syrup (Table 9, p. 40). Remove apples from solution and drain them well.

Hot Pack: Weigh slices and place them in a large, thick-bottom saucepan. Add 2 cups syrup, juice, or water for each 5 pounds of sliced apples. Bring to a boil and boil 5 minutes, stirring occasionally to prevent burning. Fill hot jars immediately with slices, leaving a ½-inch headspace. Add boiling liquid to cover apples, leaving a ½-inch headspace. Remove air bubbles. Wipe jar rims and adjust lids.

Process in a boiling water-bath canner.
Pints or Quarts, 20 minutes

For a pressure canner: Weighted-gauge canner, 10 pounds pressure
Dial-gauge canner, 11 pounds pressure
Pints or Quarts, 8 minutes

APPLESAUCE

Use sweet, juicy, crisp apples. If you prefer a tart flavor, use 1 to 2 pounds of a tart variety for each 3 pounds of sweet apples. See Table 10 (p. 41) for amount. See General Steps in Canning (pp. 19-20). Wash apples and remove stems. To help prevent darkening, prepare only one apple at a time. Peel, core, and slice each apple into a bowl of ascorbic acid solution (see Pretreating, p. 39). Remove apples from solution and drain well.

Hot Pack: Place apples in a large, thick-bottom pot. Add ½ cup water. Place over high heat, stirring occasionally to prevent burning. Cook until tender, about 5 to 20 minutes, depending on apple maturity and variety. Press apples through a sieve or food mill. Or, do not press apples if you prefer chunk-style sauce. If sugar is desired, add ⅛ cup sugar for each 4 cups of sauce. Taste and add more, if needed. Then, reheat sauce to boiling. Fill hot jars immediately with sauce, leaving a ½-inch headspace. Remove air bubbles. Wipe jar rims and adjust lids.

Process in a boiling water-bath canner.
Pints, 15 minutes
Quarts, 20 minutes

APRICOT HALVES OR SLICES

Use firm, well-colored, mature fruit of ideal quality for eating fresh or cooking. See Table 10 (p. 41) for amount. See General Steps in Canning (pp. 19-20). Dip apricots in boiling water for 30 to 60 seconds until skins loosen. Then, quickly dip in cold water and slip off skins. If you choose not to remove skins, wash them well. To help prevent darkening, prepare only one apricot at a time. Cut each apricot in half, remove pit, and then slice, if desired. Place pieces in an ascorbic acid solution (see Pretreating, p. 39). Apricots may be canned in syrup, apple or pineapple juice, white grape juice, or water. If canning in syrup, prepare a very light, light, or medium syrup (Table 9, p. 40). Remove apricots from solution and drain well.

Hot Pack: Place apricots in a large, thick-bottom saucepan. Cover them with syrup, juice, or water. Bring to a boil. Fill hot jars immediately with apricots, leaving a ½-inch headspace. If canning halves, place them in layers with the cut side down. Add boiling liquid to cover pieces, leaving a ½-inch headspace. Remove air bubbles. Wipe jar rims and adjust lids.

Process in a boiling water-bath canner.
Pints, 20 minutes
Quarts, 25 minutes

Raw Pack: Fill hot jars with raw apricots, leaving a ½-inch headspace. If canning halves, place them in layers with the cut side down. Add boiling water, juice, or syrup, leaving a ½-inch headspace. Remove air bubbles. Wipe jar rims and adjust lids.

Process in a boiling water-bath canner.
Pints, 25 minutes
Quarts, 30 minutes

For a pressure canner:
Weighted-gauge canner, 10 pounds pressure
Dial-gauge canner, 11 pounds pressure
Pints or Quarts, hot or raw pack, 10 minutes

BERRIES
(See Note, page 45)

Use ripe, sweet berries of uniform color. See Table 10 (p. 41) for amount. See General Steps in Canning (pp. 19-20). Wash only 1 or 2 quarts of berries at a time. Drain and remove caps and stems, if necessary. For gooseberries, snip off heads and tails with scissors. Berries may be canned in syrup, apple or pineapple juice, white grape juice, or water. If canning in syrup, prepare the type of your choice (Table 9, p. 40).

Hot Pack: (For blueberries, currants, elderberries, gooseberries, or huckleberries, only.) Boil water (1 gallon for each pound berries) in a large pot. Add berries to boiling water and boil 30 seconds. Drain. Pour ½ cup boiling syrup, juice, or water into each hot jar. Fill jars with berries, leaving a ½-inch headspace. Add more boiling liquid to cover berries, leaving a ½-inch headspace. Remove air bubbles. Wipe jar rims and adjust lids.

Process in a boiling water-bath canner.

Pints or Quarts, 15 minutes

Raw Pack: (For all berries listed below.) Pour ½ cup boiling syrup, juice, or water into each hot jar. Fill jars with raw berries, shaking down gently while filling, leaving a ½-inch headspace. Add more boiling syrup, juice, or water to cover berries, leaving a ½-inch headspace. Remove air bubbles. Wipe jar rims and adjust lids.

Process in a boiling water-bath canner.

Pints, 15 minutes

Quarts, 20 minutes

For a pressure canner:

Weighted-gauge canner, 10 pounds pressure

Dial-gauge canner, 11 pounds pressure

Pints or Quarts, hot or raw pack, 8 minutes

Note: Includes blackberries, blueberries, currants, dewberries, elderberries, gooseberries, huckleberries, loganberries, mulberries, and raspberries. Strawberries are not recommended for canning.

BERRY JUICE

Use freshly picked, ripe, sound fruit. See Table 10 (p. 41) for amount. See General Steps in Canning (pp. 19-20). Wash berries and remove the caps and stems. Avoid delay at any stage of the procedure.

Hot Pack: Crush fresh or frozen berries in a large, thick-bottom saucepan. Heat to boiling, reduce heat, and simmer until soft, about 5 to 10 minutes. Strain hot berries through a sieve and drain until cool enough to handle. Strain the collected juice through a double layer of cheesecloth or jelly bag. Discard the dry pulp. Add from ½ to 1 cup sugar, as desired for flavor, to each gallon of juice. Heat juice again to simmering. Fill hot jars with juice, leaving a ¼-inch headspace. Wipe jar rims and adjust lids.

Process in a boiling water-bath canner.

Pints or Quarts, 5 minutes

Note: The juice of such berries as blackberries and strawberries make delicious and wholesome drinks. They may be served as a drink or used in making sherbet, punch, chiffon pie, or salad dressing.

CHERRIES

Use bright, uniformly colored, mature, sweet or sour cherries that are ideal quality for eating fresh or cooking. See Table 10 (p. 41) for amount. See General Steps in Canning (pp. 19-20). Wash cherries and remove stems. Remove pits, if desired. If pitted, place cherries in an ascorbic acid solution (see Pretreating, p. 39) to prevent stem-end discoloration. If not pitted, prick skins on opposite sides of each cherry with a clean needle to prevent splitting. Cherries may be canned in syrup, apple or pineapple juice, white grape juice, or water. If canning in syrup, prepare the type of your choice (Table 9, p. 40). Remove pitted cherries from solution and drain well.

Hot Pack: Pour syrup, juice, or water into a large, thick-bottom saucepan. Use ½ cup liquid for each 4 cups of cherries. Add cherries and bring to a boil. Fill hot jars immediately with cherries, leaving a ½-inch headspace. Add boiling liquid to cover cherries, leaving a ½-inch headspace. Remove air bubbles. Wipe jar rims and adjust lids.

Process in a boiling water-bath canner.
Pints, 15 minutes
Quarts, 20 minutes

Raw Pack: Pour ½ cup boiling syrup, juice, or water into each hot jar. Add cherries to jars, shaking down gently as you add, leaving a ½-inch headspace. Add more boiling liquid to cover cherries, leaving a ½-inch headspace. Remove air bubbles. Wipe jar rims and adjust lids.

Process in a boiling water-bath canner.
Pints or Quarts, 25 minutes

For a pressure canner:
Weighted-gauge canner, 10 pounds pressure
Dial-gauge canner, 11 pounds pressure
Pints, 8 minutes
Quarts, 10 minutes

CRANBERRIES

Use good quality, ripe, uniformly colored berries. Discard any that are wrinkled or damaged. See Table 10 (p. 41) for amount. See General Steps in Canning (pp. 19-20). Wash cranberries and remove stems. Set aside to drain.

Hot Pack: Make a heavy syrup (Table 9, p. 40). Drop cranberries into boiling syrup and boil 3 minutes. Fill hot jars immediately with cranberries, leaving a ½-inch headspace. Add boiling syrup to cover berries, leaving a ½-inch headspace. Remove air bubbles. Wipe jar rims and adjust lids.

Process in a boiling water-bath canner.
Pints or Quarts, 15 minutes

FIGS

Use firm, ripe, uncracked figs. See Table 10 (p. 41) for amount. The color of a mature fig depends on the variety. Do not use overripe figs with very soft flesh. See General Steps In Canning (pp. 19-20). Wash figs thoroughly in clean water and drain. Do not peel or remove stems. Prepare a light syrup (Table 9, p. 40).

Hot Pack: Cover figs with water in a large, thick-bottom saucepan and bring to a boil. Boil 2 minutes and drain. Place figs in the syrup in a thick-bottom saucepan and boil gently for 5 minutes. Add acid to jars: 1 tablespoon bottled lemon juice per pint or 2 tablespoons per quart; or ¼ teaspoon citric acid per pint or ½ teaspoon per quart. Fill hot jars immediately with figs, leaving a ½-inch headspace. Add boiling syrup to cover figs, leaving a ½-inch headspace. Remove air bubbles. Wipe jar rims and adjust lids.

Process in a boiling water-bath canner.
Pints, 45 minutes
Quarts, 50 minutes

FRUIT COCKTAIL
Makes about 6 pints.

1½ pounds seedless green grapes
3 pounds peaches
3 pounds pears
3 cups sugar
4 cups water
1 10-ounce jar maraschino cherries

Use slightly underripe grapes and ripe but firm peaches. See General Steps In Canning (pp. 19-20). Wash grapes and remove stems. To prevent stem-end discoloration, place grapes in a large bowl of ascorbic acid solution (see Pretreating, p. 39). Dip peaches, a few at a time, into boiling water for 30 to 60 seconds to loosen skins. Then, place them in cold water and slip off skins. To help prevent darkening, prepare only one peach and one pear at a time. Cut each peach in half, remove the pit, and cut it into ½-inch cubes. Place them in the solution with the grapes. Peel, halve, and core each pear. Cut it into ½-inch cubes and place in the solution with grapes and peaches. Stir some to mix the fruit. Remove fruit from solution and drain well. Combine sugar and water in a large, thick-bottom saucepan and bring to a boil to make a syrup.

Raw Pack: Pour ½ cup boiling syrup into each hot jar. Add a few cherries. Then, carefully fill each jar with mixed fruit, leaving a ½-inch headspace. Add more boiling syrup to cover fruit, leaving a ½-inch headspace. Remove air bubbles. Wipe jar rims and adjust lids.

Process in a boiling water-bath canner.
Half-pints or Pints, 20 minutes

FRUIT PUREES
(See Note, below)

See General Steps in Canning (pp. 19-20). Wash fruit and remove stems. Peel and remove cores or pits, if necessary.

Hot Pack: Measure fruit into a large, thick-bottom saucepan, crushing slightly if desired. Add 1 cup hot water for each 4 cups of fruit. Cook over medium heat until fruit is soft, stirring frequently. Press through a sieve or food mill. Add a little sugar to pulp, if needed for flavor. Reheat to a boil or until sugar dissolves, if added. Fill hot jars immediately with puree, leaving a ¼-inch headspace. Remove air bubbles. Wipe jar rims and adjust lids.

Process in a boiling water-bath canner.
Pints or Quarts, 15 minutes
For a pressure canner:
Weighted-gauge canner, 10 pounds pressure
Dial-gauge canner, 11 pounds pressure
Pints or Quarts, 8 minutes

Note: Do not can pureed figs or tomatoes.

GRAPEFRUIT SECTIONS

Use firm, mature, sweet grapefruit of ideal quality for eating fresh. See Table 10 (p. 41) for amount. Grapefruit may be canned with orange sections, if desired. See General Steps In Canning (pp. 19-20). Wash, peel grapefruit, and remove sections. Discard the white membrane to prevent a bitter taste. Sections may be packed in syrup, citrus juice, or water. If using syrup, prepare a very light, light, or medium syrup (Table 9, p. 40). Or, heat juice or water to boiling to pack fruit.

Raw Pack: Fill hot jars with grapefruit sections, leaving a ½-inch headspace. Add boiling syrup, juice, or water to cover sections, leaving a ½-inch headspace. Remove air bubbles. Wipe jar rims and adjust lids.

Process in a boiling water-bath canner.
Pints or Quarts, 10 minutes
For a pressure canner:
Weighted-gauge canner, 10 pounds pressure
Dial-gauge canner, 11 pounds pressure
Pints or Quarts, 8 minutes

GRAPE JUICE

Use sweet, well-colored, firm, mature grapes of ideal quality for eating fresh. See Table 10 (p. 41) for amount. See General Steps in Canning (pp. 19-20). Wash grapes; remove stems and any green or spoiled grapes.

Hot Pack: Weigh grapes. Boil 1 cup water for each 5 pounds of grapes in a large, thick-bottom saucepan. Add grapes and return to a boil. Reduce heat and simmer slowly until grape skins are soft. Strain off juice and save. Press pulp through a sieve. Combine juices and discard pulp. Strain juice through a damp jelly bag or double layers of cheesecloth. Refrigerate juice for 24 to 48 hours (see Note, below). Without mixing, carefully pour off clear liquid and save; discard sediment. If desired, strain again through a paper coffee filter for a clearer juice. Pour juice into a saucepan and add sugar, if desired for taste. Heat and stir until sugar is dissolved. Continue heating until juice begins to boil, stirring occasionally. Fill hot jars immediately, leaving a ¼-inch headspace. Wipe jar rims and adjust lids.

Process in a boiling water-bath canner.
Pints or Quarts, 5 minutes
Half-gallons, 10 minutes

Note: Small tartaric acid (tartrate) crystals will form in grape juice. Let juice stand in the refrigerator overnight. Remove crystals before canning.

GRAPES

Use unripe, tight-skinned, preferably green seedless grapes harvested 2 weeks before they reach optimum eating quality. See Table 10 (p. 41) for amount. See General Steps in Canning (pp. 19-20). Wash grapes and remove stems. To prevent stem-end discoloration, place grapes in an ascorbic acid solution (see Pretreating, p. 39). Prepare a very light or light syrup (Table 9, p. 40). Remove grapes from solution and drain well.

Hot Pack: Weigh grapes. Measure 1 gallon water for each pound of grapes into a large, thick-bottom saucepan. Bring water to a boil. Add grapes to boiling water and boil 30 seconds; drain. Fill hot jars immediately with grapes, leaving a 1-inch headspace. Add boiling syrup to cover grapes, leaving a 1-inch headspace. Remove air bubbles. Wipe jar rims and adjust lids.

Process in a boiling water-bath canner.
Pints or Quarts, 10 minutes

Raw Pack: Fill hot jars with grapes, leaving a 1-inch headspace. Add boiling syrup to cover grapes, leaving a 1-inch headspace. Remove air bubbles. Wipe jar rims and adjust lids.

Process in a boiling water-bath canner.
Pints, 15 minutes
Quarts, 20 minutes

NECTARINE HALVES OR SLICES

Use ripe, mature nectarines of ideal quality for eating fresh or cooking. See Table 10 (p. 41) for amount. See General Steps in Canning (pp. 19-20). Wash nectarines. To help prevent darkening, prepare only one nectarine at a time. Cut each in half, and remove the pit. Slice each nectarine, if desired. Place halves or slices in an ascorbic acid solution (see Pretreating, p. 39). You can pack nectarines in syrup, apple or pineapple juice, white grape juice, or water. If using syrup, prepare a very light, light, or medium syrup (Table 9, p. 40). Remove nectarines from solution and drain well.

Hot Pack: Pour syrup, juice, or water into a large, thick-bottom saucepan. Add nectarines and heat to a boil. Fill hot jars immediately with nectarines, leaving a ½-inch headspace. If canning halves, pack them in layers with the cut side down. Add boiling liquid to cover nectarines, leaving a ½-inch headspace. Remove air bubbles. Wipe jar rims and adjust lids.

Process in a boiling water-bath canner.
Pints, 20 minutes
Quarts, 25 minutes

Raw Pack: Fill hot jars with nectarines, leaving a ½-inch headspace. If canning halves, pack them in layers with the cut side down. Add boiling syrup, juice, or water to cover nectarines, leaving a ½-inch headspace. Remove air bubbles. Wipe jar rims and adjust lids.

Process in a boiling water-bath canner.
Pints, 25 minutes
Quarts, 30 minutes

For a pressure canner:
Weighted-gauge canner, 10 pounds pressure
Dial-gauge canner, 11 pounds pressure
Pints or Quarts, 10 minutes

ORANGE SECTIONS

Use firm, mature, sweet oranges of ideal quality for eating fresh. See Table 10 (p. 41) for amount. The flavor of orange sections is best if they are canned with equal parts of grapefruit. Prepare grapefruit the same way as oranges. See General Steps in Canning (pp. 19-20). Wash, peel oranges, and remove sections. Discard the white membrane to prevent a bitter taste. Sections may be packed in syrup, citrus juice, or water. If using syrup, prepare a very light, light, or medium syrup (Table 9, p. 40). Or, heat juice or water to boiling to pack jars.

Raw Pack: Fill hot jars with orange sections, leaving a ½-inch headspace. Add boiling syrup, juice, or water to cover sections, leaving a ½-inch headspace. Remove air bubbles. Wipe jar rims and adjust lids.

Process in a boiling water-bath canner.
Pints or Quarts, 10 minutes

For a pressure canner:
Weighted-gauge canner, 10 pounds pressure
Dial-gauge canner, 11 pounds pressure
Pints or Quarts, 8 minutes

PEACH HALVES OR SLICES

Use ripe, mature peaches of ideal quality for eating fresh or cooking. See Table 10 (p. 41) for amount. See General Steps in Canning (pp. 19-20). Dip peaches in boiling water for 30 to 60 seconds until skins loosen. Then, dip quickly in cold water and slip off skins. To help prevent darkening, prepare only one peach at a time. Cut each peach in half and remove the pit. Cut it in slices, if desired. Place cut or sliced peach in an ascorbic acid solution (see Pretreating, p. 39). Peaches may be packed in syrup, apple or pineapple juice, white grape juice, or water. If using syrup, prepare a very light, light, or medium syrup (Table 9, p. 40). Remove peaches from solution and drain well.

Hot Pack: Pour syrup, juice, or water in a large, thick-bottom saucepan. Add peaches and bring to a boil. Fill hot jars immediately with peaches, leaving a ½-inch headspace. If canning halves, pack them in layers with cut side down. Add boiling liquid to cover peaches, leaving a ½-inch headspace. Remove air bubbles. Wipe jar rims and adjust lids.

Process in a boiling water-bath canner.
Pints, 20 minutes
Quarts, 25 minutes
For a pressure canner:
Weighted-gauge canner, 10 pounds pressure
Dial-gauge canner, 11 pounds pressure
Pints or Quarts, 10 minutes
Note: Raw packs produce poor quality canned peaches.

PEAR HALVES OR QUARTERS

Use ripe, mature fruit of ideal quality for eating fresh or cooking. See Table 10 (p. 41) for amount. See General Steps in Canning (pp. 19-20). Wash and peel pears. To help prevent darkening, prepare only one pear at a time. Cut each pear lengthwise in halves or quarters. Remove core and place pieces in an ascorbic acid solution (see Pretreating, p. 39). Pears can be packed in syrup, apple or pineapple juice, white grape juice, or water. If using syrup, prepare a very light, light, or medium syrup (Table 9, p. 40). Remove pears from solution and drain well.

Hot Pack: Pour syrup, juice, or water into a large, thick-bottom saucepan. Add pears and boil 5 minutes. Fill hot jars immediately with fruit, leaving a ½-inch headspace. Add boiling liquid to cover pears, leaving a ½-inch headspace. Remove air bubbles. Wipe jar rims and adjust lids.

Process in a boiling water-bath canner.
Pints, 20 minutes
Quarts, 25 minutes

For a pressure canner:
Weighted-gauge canner, 10 pounds pressure
Dial-gauge canner, 11 pounds pressure
Pints or Quarts, 10 minutes
Note: Raw packs produce poor quality canned pears.

JAPANESE PERSIMMON SAUCE

Use non-astringent variety of persimmons with crisp, sweet flesh. They should be soft-ripe. See General Steps in Canning (pp. 19-20). Wash and peel persimmons and cut into sections. Press fruit through a sieve to make a pulp.

Hot Pack: Measure pulp into a large, thick-bottom saucepan. Add ½ cup orange juice for each quart of pulp. Bring to a boil; cook until thick. Pulp will darken. Add 2 cups sugar for each quart of pulp; stir to dissolve sugar. Cook to desired consistency. Fill hot jars immediately with sauce, leaving a ½-inch headspace. Remove air bubbles. Wipe jar rims and adjust lids.

Process in a boiling water-bath canner.
Pints, 5 minutes
Quarts, 10 minutes

NATIVE PERSIMMON SAUCE

Use only ripe persimmons. See General Steps in Canning (pp. 19-20). Wash persimmons. Steam them until soft or cut them into sections. Press fruit through a sieve, food mill, or use a blender to make a pulp.

Hot Pack: Measure pulp into a large, thick-bottom saucepan. Add ½ teaspoon baking soda for each cup of pulp. Mix well to counteract the astringency of the fruit. Add about ½ cup sugar for each cup of pulp. Heat to boiling. Fill hot jars immediately with sauce, leaving a ½-inch headspace. Remove air bubbles. Wipe jar rims and adjust lids.

Process in a boiling water-bath canner.
Pints, 15 minutes
Quarts, 20 minutes

PINEAPPLE SLICES OR CUBES

Use firm, ripe pineapples. See Table 10 (p. 41) for amount. See General Steps in Canning (pp. 19-20). Wash pineapples. Peel each and remove eyes and tough fiber. Cut into slices or cubes. Pineapple may be packed in syrup, apple or pineapple juice, white grape juice, or water. If using syrup, prepare a very light, light, or medium syrup (Table 9, p. 40).

Hot Pack: Pour syrup, juice, or water into a large, thick-bottom saucepan. Add pineapple and bring to a boil. Reduce heat and simmer 10 minutes. Fill

hot jars immediately with pineapple, leaving a ½-inch headspace. Add boiling liquid to cover fruit, leaving a ½-inch headspace. Remove air bubbles. Wipe jar rims and adjust lids.
Process in a boiling water-bath canner.
Pints, 15 minutes
Quarts, 20 minutes

PLUMS

Use deep-colored, mature fruit of ideal quality for eating fresh. See Table 10 (p. 41) for amount. See General Steps in Canning (pp. 19-20). Wash plums and remove stems. To can whole, prick skins on two sides of plums with a fork to prevent splitting. Freestone varieties may be halved. Plums may be packed in syrup or water. If using syrup, prepare a very light, light, or medium syrup (Table 9, p. 40).

Hot Pack: Pour syrup or water into a large, thick-bottom saucepan. Add plums and boil 2 minutes. Remove from heat, cover saucepan, and let plums plump for 20 to 30 minutes. Then, heat to boiling. Fill hot jars immediately with plums, leaving a ½-inch headspace. Add boiling liquid to cover plums, leaving a ½-inch headspace.

Raw Pack: Fill hot jars with raw plums, packing firmly and leaving a ½-inch headspace. Add boiling syrup or water to cover plums, leaving a ½-inch headspace. Remove air bubbles. Wipe jar rims and adjust lids.

Process in a boiling water-bath canner.
Pints, 20 minutes
Quarts, 25 minutes
For a pressure canner:
Weighted-gauge canner, 10 pounds pressure
Dial-gauge canner, 11 pounds pressure
Pints or Quarts, 10 minutes

RHUBARB

Use young, tender, well-colored stalks from the spring or late fall crop. See Table 10 (p. 41) for amount. See General Steps in Canning (pp. 19-20). Trim off leaves. Wash stalks and cut into ½- to 1-inch pieces.

Hot Pack: Measure pieces into a large, thick-bottom saucepan. Add ½ cup sugar for each quart of rhubarb. Let stand until juice appears. Heat gently to boiling. Fill hot jars immediately with rhubarb, leaving a ½-inch headspace. Add boiling juice to cover rhubarb, leaving a ½-inch headspace. Remove air bubbles. Wipe jar rims and adjust lids.

Process in a boiling water-bath canner.
Pints or Quarts, 15 minutes
For a pressure canner:
Weighted-gauge canner, 10 pounds pressure
Dial-gauge canner, 11 pounds pressure
Pints or Quarts, 8 minutes

PIE FILLING RECIPES

The following fruit fillings make excellent pies. They may also be used as toppings on dessert or pastries. Each canned quart makes one 8- or 9-inch pie.

Clear Jel is a chemically modified corn starch that produces a good sauce consistency even after fillings are canned and then baked. It is a commercial product and is listed by its trade name in this book only because there were no other suitable products available to the general public at the time this book was published.

Because the different varieties of a single type of fruit may have somewhat different flavors, it may be wise to first can a single quart and make a pie with it. Then, you can adjust the sugar and spices in the recipe to suit your personal preference. However, **the proportion of lemon juice to other ingredients called for in a recipe should not be altered** because it helps control the safety and storage stability of the fillings.

APPLE PIE FILLING
Makes 7 quarts.

1 gallon water	1 teaspoon ground nutmeg (optional)
18 to 20 pounds fresh-apples	2½ cups cold water
5½ cups sugar	5 cups apple juice
1½ cups Clear Jel	7 drops yellow food coloring (optional)
1 tablespoon ground cinnamon	¾ cup bottled lemon juice

Use only fresh, firm, crisp apples. Stayman, Golden Delicious, Rome, and varieties of similar quality are suitable. See Table 10 (p. 41) for amount. If apples are not tart, use an additional 2 teaspoons of lemon juice for each quart of slices. See General Steps in Canning (pp. 19-20). Wash apples. To help prevent darkening, prepare only one apple at a time. Peel and core each apple. Slice it into ½-inch slices and place them in an ascorbic acid solution (see Pretreating, p. 39). Prepare enough apples to make 6 quarts of slices.

Hot Pack: In a large pot, boil 1 gallon of water. Place 6 cups of apples at a time into the boiling water. Allow water to return to a boil and blanch each batch 1 minute. Drain hot fruit and keep it in a covered bowl or pot while you blanch the remaining apple slices. Combine sugar, Clear Jel, cinnamon, and nutmeg in a large, thick-bottom pot. Add 2½ cups cold water, apple juice, and food coloring. Stir and cook on medium-high heat until mixture thickens and begins to bubble. Add lemon juice and boil 1 minute, stirring constantly. Quickly fold in drained apple slices. Fill hot jars immediately with pie filling, leaving a 1-inch headspace. Remove air bubbles. Wipe jar rims and adjust lids.

Process in a boiling water-bath canner.

Pints or Quarts, 25 minutes

Note: To make 1 quart, use 3½ cups apple slices, ¾ cup + 2 tablespoons sugar, ¼ cup Clear Jel, ½ teaspoon cinnamon, ⅛ teaspoon nutmeg, ½ cup cold water, ¾ cup apple juice, 1 drop food coloring, and 2 tablespoons lemon juice.

BLUEBERRY PIE FILLING
Makes 7 quarts.

1 gallon water
6 quarts fresh blueberries
6 cups sugar
2¼ cups Clear Jel

7 cups cold water
20 drops blue food coloring (optional)
7 drops red food coloring (optional)
½ cup bottled lemon juice

Use fresh, ripe, firm blueberries. See General Steps in Canning (pp. 19-20). Wash and drain blueberries.

Hot Pack: In a large pot, boil 1 gallon of water. Place 6 cups of berries at a time in the boiling water. When water returns to a boil, blanch each batch 1 minute. Drain hot berries and keep them in a covered bowl or pot while you blanch the remaining berries. Combine sugar and Clear Jel in a large, thick-bottom pot. Add 7 cups cold water and blue and red food coloring. Cook on medium-high heat until mixture thickens and begins to bubble, stirring occasionally. Add lemon juice and boil 1 minute, stirring constantly. Quickly fold in drained berries. Fill hot jars immediately with pie filling, leaving a 1-inch headspace. Remove air bubbles. Wipe jar rims and adjust lids.

Process in a boiling water-bath canner.

Pints or Quarts, 30 minutes

Note: To make 1 quart, use 3½ cups berries, ¾ cup + 2 tablespoons sugar, ¼ cup + 1 tablespoon Clear Jel, 1 cup cold water, 3 drops blue food coloring, 1 drop red food coloring, and 3½ teaspoons lemon juice.

Frozen blueberries can be used if you make the following recipe adjustments. If berries have been sweetened, rinse them while they are still frozen to remove sugar. Thaw berries in a tray or flat pan and save the juice that drains from the berries. Use it as part of the cold water required. Reduce the Clear Jel to 1¾ cups for 7 quarts and ¼ cup for 1 quart.

CHERRY PIE FILLING
Makes 7 quarts.

1 gallon water
6 quarts fresh sour cherries
7 cups sugar
1¾ cups Clear Jel
9⅓ cups cold water

1 teaspoon ground cinnamon
2 teaspoons almond extract (optional)
¼ teaspoon red food coloring (optional)
½ cup bottled lemon juice

Use fresh, very ripe but firm cherries. See General Steps in Canning (pp. 19-20). Rinse fresh cherries and remove pits. Keep them in cold water while working. To prevent stem-end discoloration, place pitted cherries in an ascorbic acid solution (see Pretreating, p. 39). Then, remove cherries from solution and drain well.

Hot Pack: In a large pot, boil 1 gallon water. Place 6 cups of cherries at a time in the boiling water. When water returns to a boil, blanch each batch 1 minute. Drain hot cherries well and keep them in a covered bowl or pot while you blanch the remaining cherries. Combine sugar and Clear Jel in a large, thick-bottom pot and add 9⅓ cups cold water. Add cinnamon, almond extract, and food coloring. Stir and cook over medium-high heat until mixture thickens and begins to bubble. Add lemon juice and boil 1 minute, stirring constantly. Quickly fold in cherries. Fill hot jars immediately with pie filling, leaving a 1-inch headspace. Remove air bubbles. Wipe jar rims and adjust lids.

Process in a boiling water-bath canner.

Pints or Quarts, 30 minutes

Note: To make 1 quart, use 3⅓ cups cherries, 1 cup sugar, ¼ cup + 1 tablespoon Clear Jel, 1⅓ cups cold water, ⅛ teaspoon cinnamon, ¼ teaspoon almond extract, 6 drops food coloring, and 1 tablespoon + 1 teaspoon lemon juice.

Frozen cherries can be used if you make the following recipe adjustments. If cherries have been sweetened, rinse them while they are still frozen to remove sugar. Thaw cherries in a tray or flat pan and save the juice that drains from the cherries. Use it as part of the cold water required. Reduce the Clear Jel to 1¾ cups for 7 quarts and ¼ cup for 1 quart.

GREEN TOMATO PIE FILLING
Makes 7 quarts.

4 quarts chopped green tomatoes
3 quarts peeled and chopped tart apples
1 pound dark seedless raisins
1 pound white raisins
¼ cup minced citron, lemon, or orange peel
2 cups water
2½ cups brown sugar
2½ cups sugar
½ cup vinegar (5% acidity)
1 cup bottled lemon juice
2 tablespoons ground-cinnamon
1 teaspoon ground nutmeg
1 teaspoon ground cloves

See General Steps In Canning (pp. 19-20).

Hot Pack: Combine all ingredients in a large, thick-bottom pot. Cook slowly, stirring often, until tomatoes and apples are tender and mixture is slightly thickened, about 30 to 40 minutes. Fill hot jars immediately with pie filling, leaving a ½-inch headspace. Remove air bubbles. Wipe jar rims and adjust lids.

Process in a boiling water-bath canner.

Quarts, 15 minutes

MINCEMEAT PIE FILLING
Makes 7 quarts.

16 to 18 pounds apples
2 cups finely chopped suet
5 pounds ground beef
2 pounds dark seedless-raisins
1 pound white raisins
2 quarts apple cider
2 tablespoons ground-cinnamon
2 teaspoons ground nutmeg
5 cups sugar
2 tablespoons canning salt

See General Steps in Canning (pp. 19-20). Peel, core, and quarter apples. Prepare enough to make 5 quarts.

Hot Pack: Cook suet and meat in enough water to avoid browning. Put meat, suet, and apples through a food grinder, using a medium blade. Place mixture in a large, thick-bottom pot and add remaining ingredients. Mix well. Heat to simmering point and simmer 1 hour or until slightly thickened. Stir often. Fill hot jars immediately with pie filling, leaving a 1-inch headspace. Remove air bubbles. Wipe jar rims and adjust lids.

Process in a pressure canner.

Weighted-gauge canner, 10 pounds pressure
Dial-gauge canner, 11 pounds pressure
Quarts, 90 minutes

To Vary: You can use 4 pounds ground venison and 1 pound sausage in place of 5 pounds ground beef.

PEACH PIE FILLING
Makes 7 quarts.

1 gallon water
16 to 17 pounds fresh peaches
7 cups sugar
2 cups + 3 tablespoons Clear Jel
1 teaspoon ground cinnamon (optional)
5¼ cups cold water
1 teaspoon almond extract (optional)
1¾ cups bottled lemon juice

Use ripe, but firm, fresh peaches. Red Haven, Redskin, Sun High, and varieties of similar quality are suitable. See General Steps in Canning (pp. 19-20). Dip peaches in boiling water for about 30 to 60 seconds until skins loosen. Then, dip quickly in cold water and slip off skins. To help prevent darkening, prepare only one peach at a time. Cut each peach in half and remove pit; cut halves into slices ½ inch thick. Place slices in an ascorbic acid solution (see Pretreating, p. 39). Prepare enough to make 6 quarts of peach slices. Remove slices from solution and drain well.

Hot Pack: Boil 1 gallon water in a large pot. Place 6 cups peach slices at a time in the boiling water. When water returns to a boil, blanch each batch 1 minute. Drain hot peaches and keep them in a covered bowl or pot while you blanch remaining slices. Combine sugar, Clear Jel, and cinnamon in a large, thick-bottom pot. Add 5¼ cups cold water and almond extract. Stir and cook over medium-high heat until mixture thickens and begins to bubble. Add lemon juice and boil 1 minute longer, stirring constantly. Fold in drained peach slices and continue to heat mixture for 3 minutes. Fill hot jars immediately with pie filling, leaving a 1-inch headspace. Remove air bubbles. Wipe jar rims and adjust lids.

Process in a boiling water-bath canner.

Pints or Quarts, 30 minutes

Note: To make 1 quart, use 3½ cups peach slices, 1 cup sugar, ¼ cup + 1 tablespoon Clear Jel, ⅛ teaspoon cinnamon, ¾ cup cold water, ⅛ teaspoon almond extract, and ¼ cup lemon juice.

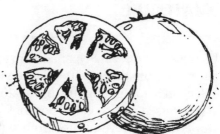

CANNING TOMATOES AND TOMATO PRODUCTS

Selecting

Use firm, vine-ripened tomatoes for canning. However, tomatoes may be canned while still green for green-tomato recipes. Do not can tomatoes from dead or frost-killed vines. Discard any that are insect-damaged, diseased, discolored, or severely bruised. Small spots of damage on otherwise healthy tomatoes can be trimmed away.

Canning

Most tomatoes are classified as an acid food and are commonly canned in a boiling water-bath canner. However, some tomato varieties have been discovered to have a low acid content and cannot be safely canned in a boiling water bath unless they are acidified with lemon juice, citric acid, or vinegar (see Adding Acid, below).

Identifying the low-acid tomatoes would not be easy for the average person; therefore, USDA recommends that all tomato products be acidified if they are to be canned in a water-bath canner. Using a pressure canner eliminates the need to acidify tomato products. Follow the manufacturer's instructions that came with your pressure canner for canning tomatoes.

The tomato recipes in this book give instructions for using a water-bath canner with the processing time for pressure canning as an option so that you can make your choice. There are a few recipes, however, which specify only one type of canner because of their other ingredients.

Adding Acid

To ensure safe acidity in whole, crushed, or juiced tomatoes, add 1 tablespoon bottled lemon juice or ¼ teaspoon citric acid to each pint of tomatoes. For quarts, use 2 tablespoons bottled lemon juice or ½ teaspoon citric acid. Acid can be added directly to the jars before filling with tomato mixture. If the result is too acid for your taste, add a little sugar. Two tablespoons vinegar (5% acidity) per pint or 4 tablespoons per quart may be used instead of lemon juice or citric acid. However, vinegar may cause undesirable flavor changes.

Caution: Heat all home-canned foods to boiling and boil 10 minutes before serving. Store unused portions in the refrigerator and use within 2 days for best quality.

TOMATO RECIPES

TOMATO JUICE

Use disease-free, vine-ripened, firm tomatoes. Discard any that have rotten spots or are from dead or frost-killed vines. See Table 10 (p. 41) for amount. See General Steps in Canning (pp. 19-20). Wash tomatoes, remove stems, and trim off bruised or discolored portions.

Hot Pack: To prevent juice from separating, quickly cut about 1 pound of tomatoes into quarters and put them directly into a large, thick-bottom saucepan. Heat immediately to boiling while crushing. Continue to cut only 1 pound at a time, add to boiling mixture, heat, and crush. Make sure the mixture boils constantly and vigorously while you add the remaining tomatoes. Simmer 5 minutes after all pieces are added.

Or, if you are not concerned about juice separation, simply slice or quarter tomatoes into a large saucepan. Crush, heat, and simmer for 5 minutes.

Press hot tomatoes through a sieve or food mill to remove skins and seeds. Heat juice again to boiling. Add acid to hot jars: 1 tablespoon bottled lemon juice or ¼ teaspoon citric acid per pint; 2 tablespoons lemon juice or ½ teaspoon citric acid per quart. Then, add canning salt, if desired: ½ teaspoon per pint or 1 teaspoon per quart. Fill hot jars immediately with juice, leaving a ½-inch headspace. Wipe jar rims and adjust lids.

Process in a boiling water-bath canner.
Pints, 35 minutes
Quarts, 40 minutes

For a pressure canner:
Weighted-gauge canner, 10 pounds pressure
Dial-gauge canner, 11 pounds pressure
Pints or Quarts, 15 minutes

TOMATO VEGETABLE JUICE
Makes about 7 quarts.

22 pounds tomatoes
3 cups finely chopped celery, onions, carrots, or peppers or any combination

Use disease-free, vine-ripened, firm tomatoes. Discard any that have rotten spots or are from dead or frost-killed vines. See General Steps in Canning (pp. 19-20). Wash tomatoes, remove stems, and trim off bruised or discolored portions.

Hot Pack: To prevent juice from separating, quickly cut about 1 pound of tomatoes into quarters and put them directly into a large, thick-bottom saucepan. Heat immediately to boiling while crushing. Continue to cut only 1 pound at a time, add to boiling mixture, heat, and crush. Make sure the mixture boils constantly and vigorously while you add the remaining tomatoes.

Or, if you are not concerned about juice separation, simply slice or quarter tomatoes into a large saucepan and crush them.

Add 3 cups of chopped vegetables to tomatoes. Heat to boiling, reduce heat, and simmer 20 minutes, stirring occasionally. Press hot tomato mixture through a sieve or food mill to remove strings, skins, and seeds. Heat juice again to boiling. Add acid to hot jars: 1 tablespoon bottled lemon juice or ¼ teaspoon citric acid per pint; 2 tablespoons lemon juice or ½ teaspoon citric acid per quart. Then, add canning salt, if desired: ½ teaspoon per pint or 1 teaspoon per quart. Fill hot jars immediately with juice, leaving a ½-inch headspace. Wipe jar rims and adjust lids.

Process in a boiling water-bath canner.
Pints, 35 minutes
Quarts, 40 minutes
For a pressure canner:
Weighted-gauge canner, 10 pounds pressure
Dial-gauge canner, 11 pounds pressure
Pints or Quarts, 15 minutes

Caution: Do not use more than 3 cups of the other vegetables for 22 pounds of tomatoes.

TOMATOES (whole or halves)

Use disease-free, vine-ripened, firm tomatoes. Discard any that have rotten spots or are from dead or frost-killed vines. See Table 10 (p. 41) for amount. See General Steps in Canning (pp. 19-20). Wash tomatoes and dip them in boiling water for 30 to 60 seconds or until skins split. Then, dip them in cold water, slip off skins, and remove cores. Leave tomatoes whole or cut them in half. *(This method is usually used when canning for fairs and exhibits.)*

Hot Pack: Put tomatoes in a large, thick-bottom saucepan and add enough water to cover them. Heat to a boil and boil gently for 5 minutes. Add acid to jars: 1 tablespoon bottled lemon juice or ¼ teaspoon citric acid per pint; 2 tablespoons lemon juice or ½ teaspoon citric acid per quart. Then, add canning salt to jars, if desired: ½ teaspoon per pint or 1 teaspoon per quart. Fill hot jars immediately with tomatoes, leaving a ½-inch headspace. Add boiling cooking liquid to cover tomatoes, leaving a ½-inch headspace.

Raw Pack: Heat water to a boil in a saucepan. Add acid to jars: 1 tablespoon bottled lemon juice or ¼ teaspoon citric acid per pint; 2 tablespoons lemon juice or ½ teaspoon citric acid per quart. Then, add canning salt to jars, if desired: ½ teaspoon per pint or 1 teaspoon per quart. Fill hot jars with raw peeled tomatoes, leaving a ½-inch headspace. Add boiling water to cover tomatoes, leaving a ½-inch headspace.

Remove air bubbles. Wipe jar rims and adjust lids.
Process in a boiling water-bath canner.
Pints, 35 minutes
Quarts, 40 minutes
For a pressure canner:
Weighted-gauge canner, 10 pounds pressure
Dial-gauge canner, 11 pounds pressure
Pints and Quarts, 10 minutes

Note: You can use heated **commercially canned tomato juice** in place of water to pack tomatoes. Follow instructions above but change processing times **to 85 minutes for pints or quarts in a boiling water-bath canner and 25 minutes in a pressure canner.**

CRUSHED TOMATOES

Use disease-free, vine-ripened, firm tomatoes. Discard any that have rotten spots or are from dead or frost-killed vines. See Table 10 (p. 41) for amount. See General Steps in Canning (pp. 19-20). Wash tomatoes and dip them in boiling water for 30 to 60 seconds or until skins split. Then, dip in cold water, slip off skins, and remove cores. Trim off any bruised or discolored portions and cut tomatoes into quarters. Divide tomato quarters into six approximately equal groups.

Hot Pack: In a large, thick-bottom saucepan, quickly heat one-sixth of the tomato quarters, crushing them with a wooden mallet or spoon as you add them. This will help draw out the juice. Continue heating the tomatoes to a boil, stirring to prevent scorching. Once tomatoes are boiling, gradually add remaining tomato quarters, stirring constantly. These tomatoes do not need to be crushed. They will soften with continued heating and stirring. Continue until all tomatoes are added. Then, boil gently for 5 minutes, stirring constantly. Add acid to jars: 1 tablespoon bottled lemon juice or ¼ teaspoon citric acid per pint; 2 tablespoons lemon juice or ½ teaspoon citric acid per quart. Then, add canning salt to jars, if desired: ½ teaspoon per pint or 1 teaspoon per quart. Fill hot jars immediately with tomatoes, leaving a ½-inch headspace. Remove air bubbles. Wipe jar rims and adjust lids.

Process in a boiling water-bath canner.
Pints, 35 minutes
Quarts, 45 minutes
For a pressure canner:
Weighted-gauge canner, 10 pounds pressure
Dial-gauge canner, 11 pounds pressure
Pints or Quarts, 15 minutes

STEWED TOMATOES
Makes about 7 quarts.

6 pounds tomatoes
¼ cup chopped green peppers
¼ cup chopped onions
2 teaspoons celery salt
2 teaspoons sugar
¼ teaspoon canning salt

Use disease-free, vine-ripened, firm tomatoes. Discard any that have rotten spots or are from dead or frost-killed vines. See General Steps in Canning (pp. 19-20). Wash tomatoes and dip them in boiling water for 30 to 60 seconds or until skins split. Then, dip them in cold water, slip off skins, and remove cores. Cut tomatoes into small pieces. Prepare enough to make about 2 quarts.

Hot Pack: Combine all ingredients in a large, thick-bottom saucepan. Cover and cook 10 minutes, stirring occasionally to prevent sticking. Fill hot jars immediately with mixture, leaving a ½-inch headspace. Remove air bubbles. Wipe jar rims and adjust lids.

Process in a pressure canner.
Weighted-gauge canner, 10 pounds pressure
Dial-gauge canner, 11 pounds pressure
Pints, 15 minutes
Quarts, 20 minutes

TOMATOES WITH OKRA OR ZUCCHINI
Makes about 7 quarts.

12 pounds tomatoes
4 pounds okra or zucchini
35 pearl onions

Use disease-free, vine-ripened, firm tomatoes. Discard any that have rotten spots or are from dead or frost-killed vines. See General Steps in Canning (pp. 19-20). Wash tomatoes and okra or zucchini. Dip tomatoes in boiling water for 30 to 60 seconds or until skins split. Then, dip them in cold water, slip off skins, and remove cores. Quarter tomatoes. Trim stems from okra and slice into 1-inch pieces or leave whole. Slice or cube zucchini, if used.

Hot Pack: Place tomatoes in a large, thick-bottom saucepan. Heat to a boil, stirring to prevent scorching. Reduce heat and simmer 10 minutes. Add okra or zucchini and boil gently 5 minutes. Add canning salt to jars, if desired: ½ teaspoon per pint or 1 teaspoon per quart. Place 5 pearl onions in each hot jar. Fill jars with tomato mixture, leaving a 1-inch headspace. Remove air bubbles. Wipe jar rims and adjust lids.

Process in a pressure canner.
Weighted-gauge canner, 10 pounds pressure
Dial-gauge canner, 11 pounds pressure
Pints, 30 minutes
Quarts, 35 minutes

Note: You can use 2 onion slices in each jar in place of 5 pearl onions.

STANDARD TOMATO SAUCE

Use disease-free, vine-ripened, firm tomatoes. Discard any that have rotten spots or are from dead or frost-killed vines. See Table 10 (p. 41) for amount. See General Steps in Canning (pp. 19-20). Wash tomatoes, remove stems, and trim off bruised or discolored portions.

Hot Pack: To prevent sauce from separating, quickly quarter about 1 pound tomatoes directly into a large, thick-bottom saucepan. Crush, heat, and simmer for 5 minutes. Heat immediately to boiling while crushing. Continue to cut only 1 pound at a time, add to boiling mixture, heat, and crush. Make sure the mixture boils constantly and vigorously while you add the remaining tomatoes. Simmer 5 minutes after all pieces are added.

Or, if you are not concerned about separation, simply slice or quarter tomatoes into a large saucepan. Crush, heat, and simmer for 5 minutes.

Press hot tomatoes through a sieve or food mill to remove skins and seeds. Pour juice into a large-diameter saucepan and simmer until it thickens. Then, boil until volume is reduced by about one-third for thin sauce or by one-half for thick sauce, stirring to prevent scorching. Add acid to jars: 1 tablespoon bottled lemon juice or ¼ teaspoon citric acid per pint; 2 tablespoons lemon juice or ½ teaspoon citric acid per quart. Then, add canning salt to jars, if desired: ½ teaspoon per pint or 1 teaspoon per quart. Fill hot jars immediately with sauce, leaving a ¼-inch headspace. Wipe jar rims and adjust lids.

Process in a boiling water-bath canner.

Pints, 35 minutes

Quarts, 40 minutes

For a pressure canner:

Weighted-gauge canner, 10 pounds pressure

Dial-gauge canner, 11 pounds pressure

Pints or Quarts, 15 minutes

SEASONED TOMATO SAUCE
Makes about 2 or 3 pints.

10 pounds tomatoes
3 medium onions
3 garlic cloves, minced
1½ teaspoons oregano
2 bay leaves

1 teaspoon canning salt
1 teaspoon black pepper
½ teaspoon crushed red pepper
1 teaspoon sugar

Use disease-free, vine-ripened, firm tomatoes. Discard any that have rotten spots or are from dead or frost-killed vines. See General Steps in Canning (pp. 19-20). Wash tomatoes and dip them in boiling water for 30 to 60 seconds or until skins split. Then, dip them in cold water, slip off skins, and remove cores. Cut tomatoes into small pieces. Peel and finely chop onions.

Hot Pack: Combine all ingredients in a large, thick-bottom saucepan. Bring to a boil, reduce heat, and simmer for 2 hours. Stir occasionally to prevent sticking. Press mixture through a food mill to remove seeds. Then, return to heat and cook over medium-high heat, stirring frequently, until thick. Fill hot jars immediately with sauce, leaving a ½-inch headspace. Wipe jar rims and adjust lids.

Process in a boiling water-bath canner.
Pints, 35 minutes

For a pressure canner:
Weighted-gauge canner, 10 pounds pressure
Dial-gauge canner, 11 pounds pressure
Pints, 15 minutes

MEXICAN TOMATO SAUCE
See Caution, below. Makes about 7 quarts.

2½ to 3 pounds chili peppers
18 pounds tomatoes
3 cups chopped onions
1 tablespoon canning salt
1 tablespoon oregano
½ cup vinegar (5% acidity)

See General Steps in Canning (pp. 19-20). Wash and dry peppers. Slit each pepper on its side to allow steam to escape. Then, blister peppers, using the Oven or Range-Top Method.

Oven or Broiler Method: Place peppers on a baking sheet or flat pan in a hot oven (400°F) or broiler for 6 to 8 minutes until skins blister.

Range-Top Method: Cover hot burner, either gas or electric, with heavy wire mesh. Place peppers on mesh for several minutes until skins blister.

Allow peppers to cool; then, place them in a pan and cover with a damp cloth. This will make peeling the peppers easier. After several minutes, peel each pepper. Discard seeds and chop peppers. Wash tomatoes and dip them in boiling water for 30 to 60 seconds or until skins split. Then, dip them in cold water, slip off skins, and remove cores. Coarsely chop tomatoes.

Hot Pack: In a large, thick-bottom saucepan, combine tomatoes with chopped peppers, onions, canning salt, oregano, and vinegar. Bring to a boil. Cover, reduce heat, and simmer for 10 minutes. Fill hot jars immediately with sauce, leaving a 1-inch headspace. Wipe jar rims and adjust lids.

Process in a pressure canner.
Weighted-gauge canner, 10 pounds pressure
Dial-gauge canner, 11 pounds pressure
Pints, 20 minutes
Quarts, 25 minutes

Caution: Wear rubber gloves while handling chilies and wash hands thoroughly with soap and water before touching your face.

TOMATO PASTE
Makes about 9 half-pints.

14 pounds tomatoes
3 sweet red peppers
2 bay leaves
1 teaspoon canning salt
1 garlic clove (optional)

Use disease-free, vine-ripened, firm tomatoes. Discard any that have rotten spots or are from dead or frost-killed vines. See General Steps in Canning (pp. 19-20). Wash tomatoes and dip them in boiling water for 30 to 60 seconds or until skins split. Then, dip them in cold water, slip off skins, and remove cores. Cut tomatoes into small pieces. Prepare enough to make about 8 quarts. Wash and drain peppers. Remove seeds and chop.

Hot Pack: In a large, thick-bottom saucepan, combine tomatoes, peppers, bay leaves, and salt. Cook over medium heat for about 1 hour. Then, press mixture through a fine sieve. Return to saucepan, add garlic, and cook slowly about 2½ more hours, stirring frequently. When sauce is thick enough to round up in a spoon, remove garlic and bay leaves. Fill hot jars immediately with paste, leaving a ½-inch headspace. Remove air bubbles. Wipe jar rims and adjust lids.

Process in a boiling water-bath canner.
Half-pints, 45 minutes

TOMATO KETCHUP
Makes 6 to 7 pints.

24 pounds ripe tomatoes
3 cups chopped onions
¾ teaspoon ground red pepper (cayenne)
3 cups cider vinegar (5% acidity)
4 teaspoons whole cloves
3 cinnamon sticks, crushed
1½ teaspoons whole allspice
3 tablespoons celery seed
1½ cups sugar
¼ cup canning salt

Use disease-free, vine-ripened, firm tomatoes. Discard any that have rotten spots or are from dead or frost-killed vines. See General Steps in Canning (pp. 19-20). Wash tomatoes and dip them in boiling water for 30 to 60 seconds or until skins split. Then, dip them in cold water, slip off skins, and remove cores.

Hot Pack: Quarter tomatoes and place them in a 4-gallon pot. Add onions and red pepper. Bring to boil, reduce heat, and simmer 20 minutes, uncovered. Pour vinegar into a 2-quart saucepan. Place cloves, cinnamon, allspice, and celery seed in a spice bag and add to vinegar. Bring to boil. Turn off heat and let stand. When tomato mixture has cooked 20 minutes, remove spice bag from vinegar and pour vinegar into tomato mixture. Stir and boil about 30 minutes. Put boiled mixture through a sieve or food mill.

Then, return tomato juice to the pot. Add sugar and salt, boil gently, and stir frequently until volume is reduced by one-half or until mixture rounds up on a spoon with no separation of liquid and solids. Fill hot jars immediately with ketchup, leaving a ⅛-inch headspace. Remove air bubbles. Wipe jar rims and adjust lids.

Process in a boiling water-bath canner.
Pints, 15 minutes.

COUNTRY WESTERN KETCHUP
See Caution (p. 68). Makes 6 to 7 pints.

5 chili peppers
24 pounds ripe tomatoes
2⅔ cups vinegar (5% acidity)
½ teaspoon ground red pepper (cayenne)
4 teaspoons paprika
4 teaspoons whole allspice
4 teaspoons dry mustard
1 tablespoon whole peppercorns
1 teaspoon mustard seed
1 tablespoon crushed bay leaves
1¼ cups sugar
¼ cup canning salt

Use disease-free, vine-ripened, firm tomatoes. Discard any that have rotten spots or are from dead or frost-killed vines. See General Steps in Canning (pp. 19-20). Wash and dry peppers. Slit each pepper on its side to allow steam to escape. Then, blister peppers, using the Oven or Range-Top Method.

Oven or Broiler Method: Place peppers on a baking sheet or flat pan in a hot oven (400°F) or broiler for 6 to 8 minutes until skins blister.

Range-Top Method: Cover hot burner, either gas or electric, with heavy wire mesh. Place peppers on mesh for several minutes until skins blister.

Allow peppers to cool; then, place them in a pan and cover with a damp cloth. This will make peeling the peppers easier. After several minutes, peel each pepper. Discard seeds and chop peppers. Wash tomatoes and dip them in boiling water for 30 to 60 seconds or until skins split. Then, dip them in cold water, slip off skins, and remove cores. Quarter tomatoes.

Hot Pack: Combine tomatoes and peppers in a 4-gallon pot. Bring to boil, reduce heat, and simmer 20 minutes, uncovered. Stir constantly to prevent sticking. Pour vinegar into a 2-quart saucepan. Combine red pepper, paprika, allspice, mustard, peppercorns, mustard seed, and bay leaves in a spice bag and add to vinegar. Bring to a boil. Turn off heat and let stand. When tomato mixture has cooked 20 minutes, remove spice bag from vinegar and pour vinegar into tomato mixture. Stir and boil about 30 minutes. Put boiled mixture through a sieve or food mill. Then, return tomato juice to the pot. Add sugar and salt, boil gently, and stir frequently until volume is reduced by one-half or until mixture rounds up on a spoon with no separation of liquid and solids. Fill hot jars immediately with ketchup, leaving a ⅛-inch headspace. Remove air bubbles. Wipe jar rims and adjust lids.

Process in a boiling water-bath canner.
Pints, 15 minutes

Caution: Wear rubber gloves while handling chilies and wash hands thoroughly with soap and water before touching your face.

BLENDER KETCHUP
Makes about 9 pints.

24 pounds ripe tomatoes	¼ cup canning salt
1 pound sweet red peppers	3 tablespoons dry mustard
1 pound sweet green peppers	1½ tablespoons ground red pepper
2 pounds onions	1½ teaspoons whole allspice
9 cups vinegar (5% acidity)	1½ tablespoons whole cloves
9 cups sugar	3 cinnamon sticks, crushed

Use disease-free, vine-ripened, firm tomatoes. Discard any that have rotten spots or are from dead or frost-killed vines. See General Steps in Canning (pp. 19-20). Wash tomatoes and dip them in boiling water for 30 to 60 seconds or until skins split. Then, dip them in cold water, slip off skins, and remove cores. Quarter tomatoes. Wash red and green peppers. Cut in half and remove seeds; then, slice into strips. Peel and quarter onions. Blend tomatoes, peppers, and onions at high speed for 5 seconds in electric blender.

Hot Pack: Pour mixture into a 3- to 4-gallon thick-bottom pot. Heat to a gentle boil and boil for 60 minutes, stirring frequently. Add vinegar, sugar, and salt. Combine dry mustard, red pepper, allspice, cloves, and cinnamon in a spice bag and add to mixture. Continue boiling and stirring until volume is reduced by one-half or until mixture rounds up on a spoon with no separation of liquid and solids. Remove spice bag. Fill hot jars immediately with ketchup, leaving a ⅛-inch headspace. Remove air bubbles. Wipe jar rims and adjust lids.

Process in a boiling water-bath canner.
Pints, 15 minutes.

Note: Using an electric blender eliminates the need for pressing or sieving.

MEATLESS SPAGHETTI SAUCE
See Caution (p. 69). Makes about 9 pints.

30 pounds tomatoes	4½ tablespoons canning salt
¼ cup vegetable oil	2 tablespoons oregano
1 cup chopped onions	4 tablespoons minced parsley
5 garlic cloves, minced	2 teaspoons black pepper
1 cup chopped celery or green pepper	¼ cup brown sugar
1 pound fresh mushrooms, sliced (optional)	

Use disease-free, vine-ripened, firm tomatoes. Discard any that have rotten spots or are from dead or frost-killed vines. See General Steps in Canning (pp. 19-20). Wash tomatoes and dip them in boiling water for 30 to 60 seconds or until skins split. Then, dip them in cold water, slip off skins, and remove cores. Cut tomatoes into quarters.

Hot Pack: Place tomatoes in a large, thick-bottom saucepan. Heat to boiling and boil 20 minutes, uncovered. Put them through a sieve or food mill to remove seeds. Heat vegetable oil in a large skillet. Add onions, garlic, celery or peppers, and mushrooms and saute until tender. Add sauteed vegetables to tomatoes in saucepan; then, add salt, oregano, parsley, pepper, and sugar. Bring to a boil, stirring to prevent sticking. Reduce heat and simmer, uncovered, until thick enough to serve. At this time, the initial volume will have been reduced by nearly one-half. Stir frequently to avoid burning. Fill hot jars immediately with sauce, leaving a 1-inch headspace. Remove air bubbles. Wipe jar rims and adjust lids.

Process in a pressure canner.
Weighted-gauge canner, 10 pounds pressure
Dial-gauge canner, 11 pounds pressure
Pints, 20 minutes
Quarts, 25 minutes

Caution: Do not increase the proportion of onions, peppers, or mushrooms.

SPAGHETTI SAUCE WITH MEAT
See Caution (p. 70). Makes about 9 pints.

30 pounds tomatoes
2½ pounds ground beef or sausage
1 cup chopped onions
5 garlic cloves, minced
1 cup chopped celery or green pepper
1 pound fresh mushrooms, sliced (optional)
4½ tablespoons canning salt
2 tablespoons oregano
4 tablespoons minced parsley
2 teaspoons black pepper
¼ cup brown sugar

Use disease-free, vine-ripened, firm tomatoes. Discard any that have rotten spots or are from dead or frost-killed vines. See General Steps in Canning (pp. 19-20). Wash tomatoes and dip them in boiling water for 30 to 60 seconds or until skins split. Then, dip them in cold water, slip off skins, and remove cores. Cut tomatoes into quarters.

Hot Pack: Place tomatoes in a large, thick-bottom saucepan. Heat to boiling and boil 20 minutes, uncovered. Put them through a sieve or food mill to remove seeds. Brown ground beef or sausage in a large skillet and drain fat. Add onions, garlic, celery or peppers, and mushrooms to beef and saute until tender. Add meat mixture to tomatoes in saucepan; then, add salt, oregano, parsley, pepper, and sugar. Bring to a boil, stirring to prevent sticking. Reduce heat and simmer, uncovered, until thick enough to serve. At this time, the initial volume will have been reduced by nearly one-half. Stir frequently to avoid burning. Fill hot jars immediately with sauce, leaving a 1-inch headspace. Remove air bubbles. Wipe jar rims and adjust lids.

Process in a pressure canner.
Weighted-gauge canner, 10 pounds pressure
Dial-gauge canner, 11 pounds pressure
Pints, 20 minutes
Quarts, 25 minutes

Caution: Do not increase the proportion of beef or sausage, onions, peppers, or mushrooms.

CHILI SALSA
See Caution, below. Makes about 7 pints.

5 pounds tomatoes	1 cup vinegar (5% acidity)
2 pounds chili peppers	3 teaspoons canning salt
1 pound onions	½ teaspoon pepper

Use disease-free, vine-ripened, firm tomatoes. Discard any that have rotten spots or are from dead or frost-killed vines. See General Steps in Canning (pp. 19-20). Wash and dry chilies. Slit each pepper on its side to allow steam to escape. Then, blister peppers, using the Oven or Range-Top Method.

Oven or Broiler Method: Place peppers on a baking sheet or flat pan in a hot oven (400°F) or broiler for 6 to 8 minutes until skins blister.

Range-Top Method: Cover hot burner, either gas or electric, with heavy wire mesh. Place peppers on mesh for several minutes until skins blister.

Allow peppers to cool; then, place them in a pan and cover with a damp cloth. This will make peeling the peppers easier. After several minutes, peel each pepper. Discard seeds and chop peppers. Wash tomatoes and dip them in boiling water for 30 to 60 seconds or until skins split. Then, dip them in cold water, slip off skins, and remove cores.

Hot Pack: Coarsely chop tomatoes and place them in a large saucepan. Peel and chop onions. Add chopped onions and peppers to tomatoes. Then, add vinegar, salt, and pepper. Heat to boiling, reduce heat, and simmer 10 minutes. Fill hot jars immediately with salsa, leaving a ½-inch headspace. Remove air bubbles. Wipe jar rims and adjust lids.

Process in a boiling water-bath canner.
Pints, 15 minutes.

Caution: Wear rubber gloves while handling chilies and wash hands thoroughly with soap and water before touching your face.

CANNING VEGETABLES

Selecting

Use only clean, fresh, good quality vegetables that are mature. Corn, however, should have slightly immature kernels, ideal quality for eating fresh. Okra and rhubarb should be young and tender. When selecting unshelled beans or peas, look for well-filled pods; snap bean pods should be tender and crisp. Beets, carrots, and sweet potatoes should be small to medium-size; larger ones tend to be fibrous. Pumpkins and winter squash should have a hard rind and stringless, mature pulp. Peppers should be firm. Discard all vegetables that are insect-damaged, diseased, wilted, discolored, or severely bruised. Small spots of damage or disease on otherwise healthy vegetables can be trimmed away.

Canning

Vegetables are classified as a low-acid food and, consequently, must be canned in a pressure canner to prevent the growth of Clostridium botulinum, the bacteria that cause botulism.

Adding Salt

Salt seasons canned vegetables but is not necessary to ensure their safety. **Use canning salt when canning vegetables. Table salt or iodized salt may discolor the food and leave a sediment due to the iodine and fillers.** As a general rule, add ½ teaspoon salt to each pint and 1 teaspoon to each quart when packing jars with vegetables. If you want to reduce sodium, follow the canning procedures in the recipes, but do not add as much salt. Vegetables may also be canned without salt. Do not use salt substitutes in canning; they may be added when serving the canned food.

Pureeing Vegetables

Do not puree vegetables before canning because proper processing times for pureed foods have not been determined for home use. For baby foods or other needs for pureed vegetables, can the vegetables whole or in pieces; then, puree them just before serving.

Caution: Heat all home-canned foods to boiling and boil 10 minutes before serving. Store unused portions in the refrigerator and use within 2 days for best quality.

Table 11. Pounds of Fresh Vegetables Needed for Standard Canner Loads

Type and Form of Vegetable	Pounds* of Whole Vegetable in 1 Bushel	Pounds of Whole Vegetable Needed to Can: 7 quarts	9 pints	1 quart
Asparagus	30+	24½	14	3½
Dried Beans	5	3¼	¾	
Lima Beans, shelled	32	28	18	4
Snap Beans, pieces	30	14	9	2
Beets, whole, sliced	52	21	13½	3
Broccoli	25+	17½	11	2½
Carrots, sliced, diced	50	17½	11	2½
Corn, cream-style	35	20		2¼
Corn, whole-kernel	35	31½	20	4½
Cucumbers	48	12¼	8	1¾
Eggplant	33	14½	7½	2
Greens, all but spinach	18	21	13½	3
Mushrooms, whole, sliced		14½	7½	2
Okra, whole, sliced	26	11	7	1½
Garden Peas	30	31½	20	4½
Southern Peas, shelled	25	26¼	17	3¾
Southern Peas, in pods	30	31½	20	4½
Hot Peppers	25	14	9	2
Sweet Peppers	25	14	9	2
Irish Potatoes, whole, cubed	50	21	13½	3
Pumpkin, cubed		16	10	2¼
Spinach	18	28	18	4
Winter Squash++, cubed	40	16	10	2¼
Sweet Potatoes, whole	50	17½	11	2½

*Vegetables weighed as follows: beans in shells; beets and carrots without tops; corn in husks; peas in pods.
+Pounds of whole vegetable in 1 crate rather than 1 bushel.

VEGETABLE RECIPES

ASPARAGUS SPEARS OR PIECES

Use tender, tight-tipped spears, 4 to 6 inches long. See Table 11 (above) for amount. See General Steps in Canning (pp. 19-20). Wash asparagus and trim off tough scales. Break off tough stems and wash again. Cut into 1-inch pieces, if desired.

Hot Pack: Boil enough water in a large pot to cover asparagus. Add asparagus to boiling water and boil 2 or 3 minutes. Loosely fill hot jars immediately with asparagus, leaving a 1-inch headspace. Add boiling cooking liquid to cover asparagus, leaving a 1-inch headspace.

Raw Pack: Fill hot jars with raw asparagus. Pack as tightly as possible without crushing, leaving a 1-inch headspace. Add boiling water to cover asparagus, leaving a 1-inch headspace.

Add canning salt to jars, if desired, ½ teaspoon per pint or 1 teaspoon per quart. Remove air bubbles. Wipe jar rims and adjust lids.

Process in a pressure canner.
Weighted-gauge canner, 10 pounds pressure
Dial-gauge canner, 11 pounds pressure
Pints, 30 minutes
Quarts, 40 minutes

BAKED BEANS
Makes about 7 quarts.

5 pounds dried beans
4 cups water
3 tablespoons dark molasses
1 tablespoon vinegar
2 teaspoons canning salt
¾ teaspoon dry mustard
7 ¾-inch cubes of pork, ham, or bacon

Use mature, shelled, dried beans or peas. Sort and discard damaged or discolored ones. See Table 11 (p. 72) for amount. See General Steps in Canning (pp. 19-20). Place dried beans or peas in a large pot and cover with water. Soak 12 to 18 hours in a cool place. Or, cover beans with boiling water and boil 2 minutes; then, remove from heat and soak 1 hour. Drain beans.

Hot Pack: Cover beans with fresh water in a large pot. Bring to a boil and boil 30 minutes. Prepare sauce by mixing 4 cups fresh water, molasses, vinegar, salt, and mustard. Heat to boiling. Place pork in a large casserole dish or crock. Thoroughly drain beans and place into dish; add enough sauce to cover beans. Cover and bake 4 to 5 hours at 350°F. Check every hour and add water as needed. Place 1 cube of pork in each hot jar. Fill jars immediately with beans, leaving a 1-inch headspace. Remove air bubbles. Wipe jar rims and adjust lids.

Process in a pressure canner.
Weighted-gauge canner, 10 pounds pressure
Dial-gauge canner, 11 pounds pressure
Pints, 65 minutes
Quarts, 75 minutes

Note: You can use 4 cups cooking liquid from beans in place of fresh water.

DRIED BEANS OR PEAS

Use mature, shelled, dried beans or peas. Sort and discard damaged or discolored ones. See Table 11 (p. 72) for amount. See General Steps in Canning (pp. 19-20). Place dried beans or peas in a large pot and cover with water. Soak 12 to 18 hours in a cool place. Or, cover beans with boiling water and boil 2 minutes; then, remove from heat and soak 1 hour. Drain beans.

Hot Pack: Cover beans with fresh water in a large pot. Bring to a boil and boil 30 minutes. Add canning salt to jars, if desired: ½ teaspoon per pint or 1 teaspoon per quart. Fill hot jars immediately with beans or peas, leaving a

1-inch headspace. Cover with hot cooking liquid, leaving a 1-inch headspace. Remove air bubbles. Wipe jar rims and adjust lids.

Process in a pressure canner.
Weighted-gauge canner, 10 pounds pressure
Dial-gauge canner, 11 pounds pressure
Pints, 75 minutes
Quarts, 90 minutes

DRIED BEANS WITH TOMATO SAUCE
Makes about 7 quarts.

5 pounds dried beans	¼ teaspoon ground cloves
1 quart tomato juice	¼ teaspoon ground allspice
3 tablespoons sugar	¼ teaspoon mace
2 teaspoons canning salt	¼ teaspoon cayenne pepper
1 tablespoon chopped onion	7 ¾-inch cubes of pork, ham, or bacon (optional)

Use mature, shelled, dried beans or peas. Sort and discard damaged or discolored ones. See General Steps in Canning (pp. 19-20). Place dried beans or peas in a large pot and cover with water. Soak 12 to 18 hours in a cool place. Or, cover beans with boiling water and boil 2 minutes; then, remove from heat and soak 1 hour. Drain beans.

Hot Pack: Cover beans with fresh water in a large pot. Bring to a boil and boil 30 minutes. Prepare sauce by mixing tomato juice, sugar, salt, onion, cloves, allspice, mace, and pepper. Heat to boiling. Fill hot jars immediately three-fourths full with hot beans. Add a cube of pork to each jar. Add boiling sauce to cover beans, leaving a 1-inch headspace. Remove air bubbles. Wipe jar rims and adjust lids.

Process in a pressure canner.
Weighted-gauge canner, 10 pounds pressure
Dial-gauge canner, 11 pounds pressure
Pints, 65 minutes
Quarts, 75 minutes

To Vary: You can make the tomato sauce with 1 cup tomato ketchup mixed with 3 cups cooking liquid from beans. Heat to boiling before filling jars.

LIMA BEANS
(Butter or Soy)

Use well-filled pods that have no insect or disease damage. See Table 11 (p. 72) for amount. Shell beans and wash them thoroughly. See General Steps in Canning (pp. 19-20).

Hot Pack: Boil enough water in a large pot to cover beans. Add beans and boil 3 minutes. Fill hot jars loosely with beans, leaving a 1-inch headspace. Add boiling cooking liquid to cover beans, leaving a 1-inch headspace.

Raw Pack: Boil a pot of fresh water to pack jars. Fill jars with raw beans. Do not press or shake down. For small beans, leave a 1-inch headspace in pints or 1½ inch in quarts. For large beans, leave a 1-inch headspace in pints or 1¼ inch in quarts. Add boiling water to cover beans, leaving a 1-inch headspace.

Add canning salt to jars, if desired: ½ teaspoon per pint or 1 teaspoon per quart. Remove air bubbles. Wipe jar rims and adjust lids.

Process in a pressure canner.
Weighted-gauge canner, 10 pounds pressure
Dial-gauge canner, 11 pounds pressure
Pints, 40 minutes
Quarts, 50 minutes

SNAP BEANS
(Green, Italian, or Wax)

Use filled, tender, crisp pods. Discard any that are diseased or appear rusty. See Table 11 (p. 72) for amount. See General Steps in Canning (pp. 19-20). Wash beans and trim ends. Leave beans whole or cut into 1-inch pieces.

Hot Pack: Boil enough water in a large pot to cover beans. Add beans to boiling water and boil 5 minutes. Fill hot jars loosely with beans, leaving a 1-inch headspace. Add boiling cooking liquid to cover beans, leaving a 1-inch headspace.

Raw Pack: Boil a pot of water to pack jars. Fill hot jars tightly with raw beans, leaving a 1-inch headspace. Add boiling water to cover beans, leaving a 1-inch headspace.

Add canning salt to jars, if desired: ½ teaspoon per pint or 1 teaspoon per quart. Remove air bubbles. Wipe jar rims and adjust lids.

Process in a pressure canner.
Weighted-gauge canner, 10 pounds pressure
Dial-gauge canner, 11 pounds pressure
Pints, 20 minutes
Quarts, 25 minutes

BEETS

Use beets no larger than 3 inches in diameter; larger beets are often fibrous. To can whole beets, use those with a diameter of 1 to 2 inches. See Table 11 (p. 72) for amount. See General Steps in Canning (pp. 19-20). It is best to leave 1 inch of beet tops and roots during the early stage of preparation to reduce bleeding of color. Scrub beets well.

Hot Pack: Boil enough water in a large pot to cover beets. Add beets to boiling water and boil until skins slip easily; about 15 to 25 minutes, depending on size. Drain, cool, and remove skins. Trim off stems and roots. Leave baby beets whole. Slice medium or large beets or cut into ½-inch cubes. Halve or quarter very large slices. Boil a pot of fresh water to pack jars. Add canning salt to jars, if desired: ½ teaspoon per pint or 1 teaspoon per quart. Fill hot jars immediately with beets, leaving a 1-inch headspace. Add fresh boiling water to cover beets, leaving a 1-inch headspace. Remove air bubbles. Wipe jar rims and adjust lids.

Process in a pressure canner.
Weighted-gauge canner, 10 pounds pressure
Dial-gauge canner, 11 pounds pressure
Pints, 30 minutes
Quarts, 35 minutes

SLICED OR DICED CARROTS

Use carrots no larger than 1 to 1¼ inches in diameter; larger carrots are often fibrous. See Table 11 (p. 72) for amount. See General Steps in Canning (pp. 19-20). Wash and peel carrots; then, rewash them. Slice or dice them.

Hot Pack: Boil enough water in a large pot to cover carrots. Add carrots to boiling water, reduce heat, and simmer 5 minutes. Fill hot jars loosely with carrots, leaving a 1-inch headspace. Add boiling cooking liquid to cover carrots, leaving a 1-inch headspace.

Raw Pack: Boil a pot of water to pack jars. Fill hot jars tightly with raw carrots, leaving a 1-inch headspace. Add boiling water to cover carrots, leaving a 1-inch headspace.

Add canning salt to jars, if desired: ½ teaspoon per pint or 1 teaspoon per quart. Remove air bubbles. Wipe jar rims and adjust lids.

Process in a pressure canner.
Weighted-gauge canner, 10 pounds pressure
Dial-gauge canner, 11 pounds pressure
Pints, 25 minutes
Quarts, 30 minutes

CREAM-STYLE CORN

Use ears with slightly immature kernels or of ideal quality for eating fresh. See Table 11 (p. 72) for amount. See General Steps in Canning (pp. 19-20). Remove husks and silks and wash corn.

Hot Pack: Boil enough water in a large pot to cover corn. Place corn in boiling water and boil 4 minutes. Cut corn from cobs at about the center of the kernels. Then, scrape the remaining corn from cobs with a table knife. Boil a pot of fresh water. Measure corn and scrapings into a large pot. Add 1 cup boiling water for every 2 cups of corn. Heat to boiling and simmer 3 minutes, stirring constantly. Add ½ teaspoon canning salt to each pint jar, if desired. Fill hot jars immediately with corn, leaving a 1-inch headspace. Remove air bubbles. Wipe jar rims and adjust lids.

Process in a pressure canner.
Weighted-gauge canner, 10 pounds pressure
Dial-gauge canner, 11 pounds pressure
Pints, 85 minutes.

Caution: Do not use raw-pack method for cream-style corn; do not can cream-style corn in quart jars.

WHOLE-KERNEL CORN

Use ears with slightly immature kernels or of ideal quality for eating fresh. See Table 11 (p. 72) for amount. See General Steps in Canning (pp. 19-20). Remove husks and silks and wash corn.

Hot Pack: Boil enough water in a large pot to cover corn. Place corn in boiling water and boil 3 minutes. Boil a pot of fresh water. Measure corn into a large saucepan. Add 1 cup boiling water for every 4 cups of corn. Heat to boiling, reduce heat, and simmer 5 minutes. Fill hot jars immediately with corn, leaving a 1-inch headspace. Add hot cooking liquid, leaving a 1-inch headspace.

Raw Pack: Cut corn from cob at about three-fourths the depth of the kernels. **Do not scrape cob.** Boil a pot of water to pack jars. Fill hot jars with raw kernels, leaving a 1-inch headspace. Do not shake or press down. Add boiling water to cover corn, leaving a 1-inch headspace.

Add canning salt to jars, if desired: ½ teaspoon per pint or 1 teaspoon per quart. Remove air bubbles. Wipe jar rims and adjust lids.

Process in a pressure canner.
Weighted-gauge canner, 10 pounds pressure
Dial-gauge canner, 11 pounds pressure
Pints, 55 minutes
Quarts, 85 minutes

Caution: Some of the sweeter varieties or corn with kernels that are too mature may turn brown when canned. To test, can only a small amount and check the color and flavor before canning large quantities.

GREENS
(Collards, Mustard, Spinach, or Turnips)

Use only freshly harvested greens. Discard any wilted, discolored, diseased, or insect-damaged leaves. Leaves should be tender and have a good color. See Table 11 (p. 72) for amount. See General Steps in Canning (pp. 19-20). Wash only small amounts of greens at one time. Drain water and continue rinsing until water is clear and free of grit. Cut out tough stems and midribs.

Hot Pack: Boil enough water in a large pot to blanch greens. Place 1 pound of greens at a time in a cheesecloth bag or blancher basket and steam 3 to 5 minutes or until well wilted. Boil a pot of fresh water to pack jars. Add canning salt to jars, if desired: ¼ teaspoon per pint or ½ teaspoon per quart. Fill hot jars loosely with greens, leaving a 1-inch headspace. Add boiling water to cover greens, leaving a 1-inch headspace. Remove air bubbles. Wipe jar rims and adjust lids.

Process in a pressure canner.
Weighted-gauge canner, 10 pounds pressure
Dial-gauge canner, 11 pounds pressure
Pints, 70 minutes
Quarts, 90 minutes

MUSHROOMS

Use only brightly colored, small to medium-size domestic mushrooms with short stems and tight veils (unopened caps). See Table 11 (p. 72) for amount. See General Steps in Canning (pp. 19-20). Trim stems and small discolored parts. Soak mushrooms in cold water for 10 minutes to remove dirt. Wash in clean water. Leave small mushrooms whole; slice large ones. Separate whole from sliced mushrooms, if desired, for uniform packs.

Hot Pack: Place mushrooms in a large saucepan and cover with water. Heat to boiling and boil 5 minutes. Boil a pot of fresh water to pack jars. Fill hot jars immediately with mushrooms, leaving a 1-inch headspace. Add canning salt to jars, if desired: ¼ teaspoon per half-pint or ½ teaspoon per pint. For better color, add ⅛ teaspoon ascorbic acid powder per pint. Add boiling water to cover mushrooms, leaving a 1-inch headspace. Remove air bubbles. Wipe jar rims and adjust lids.

Process in a pressure canner.
Weighted-gauge canner, 10 pounds pressure
Dial-gauge canner, 11 pounds pressure
Half-pints or pints, 45 minutes
Caution: Do not can wild mushrooms.

OKRA

Use young, tender pods. Remove and discard diseased and rust-spotted ones. See Table 11 (p. 72) for amount. See General Steps in Canning (pp. 19-20). Wash pods and trim ends. Leave okra whole or cut into 1-inch pieces.

Hot Pack: Boil enough water in a large pot to cover okra. Add okra to boiling water and boil 2 minutes. Fill hot jars immediately with okra, leaving a 1-inch headspace. Add boiling cooking liquid, leaving a 1-inch headspace. Add canning salt to jars, if desired: ½ teaspoon per pint or 1 teaspoon per quart. Remove air bubbles. Wipe jar rims and adjust lids.

Process in a pressure canner.
Weighted-gauge canner, 10 pounds pressure
Dial-gauge canner, 11 pounds pressure
Pints, 25 minutes
Quarts, 40 minutes

ONIONS

Use firm, young onions of 1-inch diameter or less. Discard any that are soft or damaged. See General Steps in Canning (pp. 19-20). Wash and peel onions.

Hot Pack: Boil enough water in a large pot to cover onions. Add onions and boil 5 minutes. Boil a pot of fresh water to pack jars. Fill hot jars immediately with onions, leaving a 1-inch headspace. Add canning salt to jars, if desired: ½ teaspoon per pint or 1 teaspoon per quart. Add boiling water to cover onions, leaving a 1-inch headspace. Remove air bubbles. Wipe jar rims and adjust lids.

Process in a pressure canner.
Weighted-gauge canner, 10 pounds pressure
Dial-gauge canner, 11 pounds pressure
Pints or Quarts, 40 minutes

GARDEN PEAS
(English or Green)

Use filled pods containing young, tender, sweet peas. Discard any diseased pods. See Table 11 (p. 72) for amount. See General Steps in Canning (pp. 19-20). Shell and wash peas.

Hot Pack: Boil enough water in a large pot to cover peas. Add peas to boiling water and boil 2 minutes. Fill hot jars loosely with peas, leaving a 1-inch headspace. Add boiling cooking liquid to cover peas, leaving a 1-inch headspace.

Raw Pack: Boil a pot of water to pack jars. Fill jars with raw peas, leaving a 1-inch headspace. Do not shake or press down. Add boiling water to cover peas, leaving a 1-inch headspace.

Add canning salt to jars, if desired: ½ teaspoon per pint or 1 teaspoon per quart. Remove air bubbles. Wipe jar rims and adjust lids.

Process in a pressure canner.
Weighted-gauge canner, 10 pounds pressure
Dial-gauge canner, 11 pounds pressure
Pints or Quarts, 40 minutes

Caution: For best quality, freeze sugar snap peas and Chinese edible pods; do not can them.

SOUTHERN PEAS
(Blackeye, Crowder, or Field)

Use young, tender peas from well-filled pods. Discard diseased or damaged pods. See Table 11 (p. 72) for amount. See General Steps in Canning (pp. 19-20). Shell peas; wash and drain well.

Hot Pack: Boil enough water in a large pot to cover peas. Add peas to boiling water and boil 3 minutes. Fill hot jars loosely with peas, leaving a 1-inch headspace for pints or 1½-inch headspace for quarts. Add boiling cooking liquid to cover peas, leaving a 1-inch headspace for pints or 1½ inch for quarts.

Raw Pack: Boil a pot of water to pack jars. Fill hot jars loosely with raw peas, leaving a 1-inch headspace for pints or 1½-inch headspace for quarts. Add boiling water to cover peas, leaving a 1-inch headspace for pints or 1½ inch for quarts.

Add canning salt to jars, if desired: ½ teaspoon per pint or 1 teaspoon per quart. Remove air bubbles. Wipe jar rims and adjust lids.

Process in a pressure canner.
Weighted-gauge canner, 10 pounds pressure
Dial-gauge canner, 11 pounds pressure
Pints, 40 minutes
Quarts, 50 minutes

HOT PEPPERS
See Note and Caution (p. 81)

Use firm well-colored peppers. Do not use soft or diseased ones. See Table 11 (p. 72) for amount. See General Steps in Canning (pp. 19-20). Wash and drain peppers.

Hot Pack: Slit each pepper on its side to allow steam to escape. Then, use the Oven or Range-Top Method to roast and peel peppers.

Oven Or Broiler Method: Put peppers on a baking sheet and place in a hot oven (400°F) or broiler for 6 to 8 minutes until skins blister.

Range-Top Method: Cover a hot burner, either gas or electric, with a heavy wire mesh. Place peppers on mesh for several minutes until skins blister.

Allow peppers to cool; then, place them in a pan and cover with a damp cloth. This will make peeling the peppers easier. After several minutes, peel each pepper. Small peppers may be left whole; quarter larger ones and remove stems and seeds. Flatten whole peppers. Boil a pot of fresh water to pack jars. Add canning salt to jars, if desired: ¼ teaspoon per half-pint or ½ teaspoon per pint. Fill hot jars loosely with peppers, leaving a 1-inch headspace. Add boiling water to cover peppers, leaving a 1-inch headspace. Remove air bubbles. Wipe jar rims and adjust lids.

Process in a pressure canner.
Weighted-gauge canner, 10 pounds pressure
Dial-gauge canner, 11 pounds pressure
Half-pints and pints, 35 minutes
Note: Includes chili, Hungarian, and jalapeno peppers.
Caution: Wear rubber gloves while handling hot peppers and wash your hands thoroughly before touching your face.

PIMIENTO PEPPERS

Use firm, well-colored peppers. Discard soft or diseased ones. See Table 11 (p. 72) for amount. See General Steps in Canning (pp. 19-20). Wash and drain peppers.

Hot Pack: Boil enough water in a large, thick-bottom saucepan to cover peppers. Place peppers in boiling water and boil about 10 or 20 minutes. Or, roast them on a baking sheet in a 400°F oven about 6 to 8 minutes until the skins can be rubbed off. Remove skins, stems, blossom ends, and seeds. Flatten peppers.

Boil a pot of fresh water to pack jars. Fill hot jars immediately with peppers, leaving a 1-inch headspace. Add ½ teaspoon canning salt to pint jars, if desired. Add boiling water to cover peppers, leaving a 1-inch headspace. Remove air bubbles. Wipe jar rims and adjust lids.

Process in a pressure canner.
Weighted-gauge canner, 10 pounds pressure
Dial-gauge canner, 11 pounds pressure
Pints, 35 minutes

SWEET PEPPERS

Use firm green, red, or yellow peppers. Discard soft or diseased ones. See Table 11 (p. 72) for amount. See General Steps in Canning (pp. 19-20). Wash and drain peppers. Remove stems and seeds and quarter large peppers; small ones can be left whole.

Hot Pack: Boil enough water in a large, thick-bottom pot to cover peppers. Place pieces of pepper in boiling water and boil 3 minutes. Boil a pot of fresh water to pack jars. Fill hot jars immediately with peppers, leaving a 1-inch headspace. Add ½ teaspoon canning salt to pint jars, if desired. Add boiling water to cover peppers, leaving a 1-inch headspace. Remove air bubbles. Wipe jar rims and adjust lids.

Process in a pressure canner.
Weighted-gauge canner, 10 pounds pressure
Dial-gauge canner, 11 pounds pressure
Pints, 35 minutes

IRISH POTATOES

Use small to medium-size, mature potatoes of ideal quality for cooking. See Table 11 (p. 72) for amount. See General Steps in Canning (pp. 19-20). Wash and peel potatoes. If desired, cut them into ½-inch cubes. Use potatoes 1 to 2 inches in diameter to can whole. As each potato is peeled or cubed, place it in an ascorbic acid solution (see Pretreating, p. 39) to prevent darkening. When all potatoes are prepared, remove them from the solution and drain well.

Hot Pack: Place cubed potatoes in a large pot, cover them with water, and bring to a boil; boil 2 minutes and drain. For whole potatoes, boil 10 minutes and drain. Boil a pot of fresh water to pack jars. Fill hot jars immediately with potatoes, leaving a 1-inch headspace. Add canning salt to jars, if desired: ½ teaspoon per pint or 1 teaspoon per quart. Add boiling water to cover potatoes, leaving a 1-inch headspace. Remove air bubbles. Wipe jar rims and adjust lids.

Process in a pressure canner.
Weighted-gauge canner, 10 pounds pressure
Dial-gauge canner, 11 pounds pressure
Pints, 35 minutes
Quarts, 40 minutes

Caution: Potatoes that have been stored at 45°F or below may discolor when canned.

PUMPKIN

Use pumpkins that have a hard rind and stringless, mature pulp of ideal quality for cooking fresh. Small pumpkins (sugar or pie varieties) make better canned products. See Table 11 (p. 72) for amount. See General Steps in Canning (pp. 19-20). Wash each pumpkin and cut it in half to remove seeds; then, cut it into slices 1 inch wide. Peel slices and cut them into 1-inch cubes.

Hot Pack: Boil enough water in a large pot to cover pumpkin. Place cubes in boiling water and boil 2 minutes. Do not mash or puree the pumpkin. Fill hot jars immediately with pumpkin, leaving a 1-inch headspace. Add boiling cooking liquid to cover pieces, leaving a 1-inch headspace. Remove air bubbles. Wipe jar rims and adjust lids.

Process in a pressure canner.
Weighted-gauge canner, 10 pounds pressure
Dial-gauge canner, 11 pounds pressure
Pints, 55 minutes
Quarts, 90 minutes

WINTER SQUASH

Use squash that have a hard rind and stringless, mature pulp of ideal quality for cooking fresh. See Table 11 (p. 72) for amount. See General Steps in Canning (pp. 19-20). Wash each squash and cut it in half to remove seeds. Then, cut it into slices 1 inch wide. Peel slices and cut them into 1-inch cubes.

Hot Pack: Boil enough water in a large pot to cover squash. Add cubes to boiling water and boil 2 minutes. **Do not mash or puree squash.** Fill hot jars immediately with squash, leaving a 1-inch headspace. Add canning salt, if desired: ½ teaspoon per pint or 1 teaspoon per quart. Add boiling cooking liquid to cover squash, leaving a 1-inch headspace. Remove air bubbles. Wipe jar rims and adjust lids.

Process in a pressure canner.
Weighted-gauge canner, 10 pounds pressure
Dial-gauge canner, 11 pounds pressure
Pints, 55 minutes
Quarts, 90 minutes

Caution: Do not can summer squash; freeze them.

SUCCOTASH
Makes about 7 quarts.

12 cups whole-kernel corn (p. 77)   16 cups shelled lima beans (p. 74)
8 cups whole or crushed tomatoes (pp. 61-62) (optional)

See General Steps In Canning (pp. 19-20). Wash and prepare corn, beans, and tomatoes as described for each on the indicated pages.

Hot Pack: Combine all prepared vegetables in a large pot. Add enough water to cover vegetables. Bring to a boil, reduce heat, and simmer 5 minutes. Add canning salt to jars, if desired: ½ teaspoon per pint or 1 teaspoon per quart. Fill hot jars immediately with vegetables, leaving a 1-inch headspace. Add boiling cooking liquid to cover vegetables, leaving a 1-inch headspace.

Raw Pack: Boil a pot of water to pack jars. Fill hot jars with equal parts of all prepared vegetables, leaving a 1-inch headspace. Do not shake or press down. Add canning salt to jars, if desired: ½ teaspoon per pint or 1 teaspoon per quart. Add boiling water to cover vegetables, leaving a 1-inch headspace.

Remove air bubbles. Wipe jar rims and adjust lids.

Process in a pressure canner.
Weighted-gauge canner, 10 pounds pressure
Dial-gauge canner, 11 pounds pressure
Pints, 60 minutes
Quarts, 85 minutes

Note: You will need about 15 pounds of unhusked corn for 12 cups of whole kernels and about 14 pounds of unshelled lima beans for 16 cups shelled beans.

SWEET POTATOES

Choose small to medium-size potatoes. They should be mature but not too fibrous. See Table 11 (p. 72) for amount. Can potatoes within 1 to 2 months after harvest. See General Steps in Canning (pp. 19-20). Wash potatoes.

Hot Pack: Boil or steam potatoes until they are partially soft, about 15 to 20 minutes. Boil a pot of fresh water or make a syrup (Table 9, p. 40) to pack jars. Remove skins and cut medium potatoes, if needed, so that pieces are uniform in size. **Do not mash or puree pieces.** Fill hot jars with potatoes, leaving a 1-inch headspace. Add canning salt to jars, if desired: ½ teaspoon per pint or 1 teaspoon per quart. Add boiling syrup or water to cover potatoes, leaving a 1-inch headspace. Remove air bubbles. Wipe jar rims and adjust lids.

Process in a pressure canner.
Weighted-gauge canner, 10 pounds pressure
Dial-gauge canner, 11 pounds pressure
Pints, 65 minutes
Quarts, 90 minutes

Caution: Do not can sweet potatoes without liquid.

TURNIP ROOTS

Use fresh, firm turnips. Discard any that are diseased or damaged. See General Steps in Canning (pp. 19-20). Wash roots, scrubbing well. Peel, slice, or dice them.

Hot Pack: Boil enough water in a large pot to cover roots. Place them in boiling water and boil 5 minutes. Drain. Fill hot jars immediately with turnip pieces, leaving a 1-inch headspace. Add canning salt to jars, if desired: ½ teaspoon per pint or 1 teaspoon per quart. Add boiling cooking liquid to cover turnips, leaving a 1-inch headspace. Remove air bubbles. Wipe jar rims and adjust lids.

Process in a pressure canner.
Weighted-gauge canner, 10 pounds pressure
Dial-gauge canner, 11 pounds pressure
Pints, 30 minutes
Quarts, 35 minutes

MIXED VEGETABLES
Makes 7 quarts.

6 cups sliced carrots (p. 76)
6 cups whole-kernel corn (p. 77)
6 cups cut snap beans (p. 75)
6 cups lima beans (p. 74)
4 cups whole or crushed tomatoes (pp. 61-62)
4 cups sliced zucchini

See General Steps In Canning (pp. 19-20). Wash and prepare the carrots, corn, snap beans, lima beans, and tomatoes as described for each on the indicated pages. Wash, trim, and slice or cube zucchini.

Hot Pack: Combine all vegetables in a large pot and add enough water to cover them. Heat to boiling and boil 5 minutes. Add canning salt to jars, if desired: ½ teaspoon per pint or 1 teaspoon per quart. Fill hot jars immediately with vegetables, leaving a 1-inch headspace. Add boiling cooking liquid to cover vegetables, leaving a 1-inch headspace. Remove air bubbles. Wipe jar rims and adjust lids.

Process in a pressure canner.
Weighted-gauge canner, 10 pounds pressure
Dial-gauge canner, 11 pounds pressure
Pints, 75 minutes
Quarts, 90 minutes

To Vary: You may use more of one vegetable and less of another, if desired. Or, you may substitute any other favorite vegetable, except leafy greens, dried beans, cream-style corn, winter squash, or sweet potatoes.

VEGETABLE SOUP

Select, wash, and prepare your choice of vegetables and meat or poultry as described for the specific foods (see Index, pp. 253-265).

Hot Pack: In a large pot, cover meat or poultry with water and cook until tender. Cool and remove bones. Cook vegetables separately. If using dried beans or peas, add 3 cups of water to each cup of beans in a saucepan and boil 2 minutes; remove from heat, soak 1 hour, and heat again to boiling. Drain vegetables and save cooking liquid. Combine vegetables with meat or poultry, tomatoes, broth, or cooking liquid from vegetables. Do not thicken broth. Add canning salt to jars, if desired: ½ teaspoon per pint or 1 teaspoon per quart. Fill hot jars half-full with the solid mixture. Then, add the remaining liquid, leaving a 1-inch headspace. Remove air bubbles. Wipe jar rims and adjust lids.

Process in a pressure canner.
Weighted-gauge canner, 10 pounds pressure
Dial-gauge canner, 11 pounds pressure
Pints, 60 minutes
Quarts, 75 minutes

To Vary: You can add raw seafood after vegetables have been cooked and other ingredients are being combined. If seafood is added, process pints and quarts 100 minutes.

ZUCCHINI
Makes about 9 pints.

4 quarts cubed or shredded zucchini
46 ounces canned unsweetened pineapple juice
1½ cups bottled lemon juice
3 cups sugar

See General Steps in Canning (pp. 19-20). Peel zucchini and cut into ½-inch cubes or shred it.

Hot Pack: In a large pot, mix zucchini with pineapple juice, lemon juice, and sugar. Bring to a boil, reduce heat, and simmer 20 minutes. Fill hot jars immediately with zucchini, leaving a ½-inch headspace. Add boiling cooking liquid to cover zucchini, leaving a ½-inch headspace. Remove air bubbles. Wipe jar rims and adjust lids.

Process in a boiling water-bath canner.

Half-pints or Pints, 15 minutes

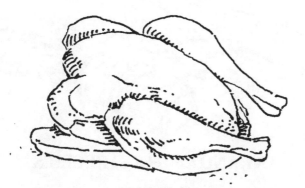

CANNING MEAT, POULTRY, AND SEAFOOD

Most kinds of meat, poultry, and seafood may be successfully canned at home, including game animals as well as domestic. Observe the following safety guidelines and quality suggestions when canning these products:

• Do not can meats in combination with other foods without specific recipes.

• Use only meat from healthy animals that were slaughtered under strict sanitary conditions.

• Chill meat and poultry at once after slaughter and keep it chilled at 40°F or lower until canning time. Dressed poultry and small animals should be chilled no less than 6 hours and no more than 12 before canning. If they must be held for a longer period, freeze them and store at a temperature of 0°F or lower.

• Keep all meat clean and sanitary as you work with it.

• Trim meat of gristle, bruised spots, and fat before canning. Fat left on meat will melt and climb the sides of the jar during processing. If fat comes in contact with the sealing edge of the lid, the jar may not seal.

• **Do not puree meat or poultry before canning.** Proper processing times for pureed foods have not been determined for home use. Instead, can pieces of meat and poultry and puree them at serving time.

• Salt seasons the food but is not necessary for safety. Salt may be omitted when canning meat, poultry, and fish. If using salt, use canning salt. Table salt may discolor the food and leave a sediment due to the iodine and fillers.

• Use a pressure canner for canning meats, poultry, and fish because they are low-acid foods.

• Follow canning procedures exactly; otherwise, the recommended processing times will not give adequate sterilization.

Caution: Heat all home-canned foods to boiling and boil 10 minutes before serving. Store unused portions in the refrigerator and use within 2 days for best quality.

MEAT, POULTRY, AND SEAFOOD RECIPES

MEAT CUTS
(Chops, Roasts, or Steaks)

Use high quality beef, lamb, pork, veal, venison, and other large wild game. See General Steps in Canning (pp. 19-20). Prepare the tender cuts of meat that are suitable for roasts, steaks, and chops as whole pieces. Can the less tender cuts and smaller pieces as stew meat. Soak strong-flavored wild meats 1 hour in salt water (1 tablespoon canning salt per quart of water); then rinse. Do not add salt to jars for presoaked meats. Remove large bones.

Hot Pack: Cook meat until done by roasting, stewing, or browning in a small amount of fat. Cooked meat may be canned in boiling broth, tomato juice, or water. Prepare broth or heat juice or water and keep it hot. Add canning salt to jars, if desired: ½ teaspoon to pints and 1 teaspoon to quarts. Fill hot jars with meat, leaving a 1-inch headspace. Add boiling broth, juice, or water, leaving a 1-inch headspace. Remove air bubbles. Wipe jar rims and adjust lids.

Raw Pack: Add canning salt to jars, if desired: ½ teaspoon to pints or 1 teaspoon to quarts. Fill hot jars with raw meat pieces, leaving a 1-inch headspace. Do not add liquid. Wipe jar rims and adjust lids.

Process in a pressure canner.
Weighted-gauge canner, 10 pounds pressure
Dial-gauge canner, 11 pounds pressure
Pints, 75 minutes
Quarts, 90 minutes

Note: Tomato juice is recommended for adding to wild game if using the hot-pack method.

GROUND OR CHOPPED MEAT

Use fresh, chilled beef, lamb, pork, veal, venison, and other large wild game. See General Steps in Canning (pp. 19-20). For venison, add one part high-quality pork fat to three or four parts venison before grinding. Use freshly made sausage, seasoned with canning salt and cayenne pepper; sage may cause a bitter off-flavor if canned.

Hot Pack: Shape chopped meat into patties or balls or cut cased sausage into 3- to 4-inch links. Cook until lightly browned. Ground meat may be sauteed without shaping. Remove excess fat. Cooked meat may be canned in boiling broth, tomato juice, or water. Prepare broth or heat juice or water. Fill hot jars with meat, leaving a 1-inch headspace. Add boiling liquid, leaving a 1-inch headspace. Add canning salt to jars, if desired: ½ teaspoon per pint or 1 teaspoon per quart. Remove air bubbles. Wipe jar rims and adjust lids.

Process in a pressure canner.
Weighted-gauge canner, 10 pounds pressure
Dial-gauge canner, 11 pounds pressure
Pints, 75 minutes
Quarts, 90 minutes

MEAT BROTH

Use fresh, trimmed bones of beef, lamb, pork, venison, or other game animals. See General Steps In Canning (pp. 19-20). Saw or crack the bones for more flavor.

Hot Pack: Rinse bones and place them in a large pot. Add enough water to cover bones. Cover pot and simmer for 3 to 4 hours. Remove bones and pick off meat. Cool broth; then, skim off and discard excess fat. Return pieces of meat to broth and heat to boiling. Fill hot jars immediately with meat and broth, leaving a 1-inch headspace. Wipe jar rims and adjust lids.

Process in a pressure canner.
Weighted-gauge canner, 10 pounds pressure
Dial-gauge canner, 11 pounds pressure
Pints, 20 minutes
Quarts, 25 minutes

POULTRY, RABBIT, AND SQUIRREL

Use freshly killed and dressed healthy animals. See General Steps In Canning (pp. 19-20). Large chickens are more flavorful than fryers. Dressed chickens should be chilled between 6 and 12 hours before canning. Soak strong-flavored game birds, dressed rabbits, and squirrels 1 hour in salt water (1 tablespoon canning salt per quart water); then rinse. Do not add salt to jars for presoaked animals. Remove excess fat. Cut animals into suitable sizes for filling jars. Can meat with or without bones.

Hot Pack: Boil, steam, or bake meat until about two-thirds done. Prepare broth and keep it hot. Add canning salt to jars, if desired: ½ teaspoon per pint or 1 teaspoon per quart. Fill hot jars with meat, leaving a 1¼-inch headspace. Add boiling broth to cover meat, leaving a 1¼-inch headspace. Remove air bubbles. Wipe jar rims and adjust lids.

Raw Pack: Add canning salt to jars, if desired: ½ teaspoon per pint or 1 teaspoon per quart. Fill hot jars loosely with raw meat, leaving a 1¼-headspace. Do not add liquid. Wipe jar rims and adjust lids.

Process in a pressure canner.
Weighted-gauge canner, 10 pounds pressure
Dial-gauge canner, 11 pounds pressure
For jars without bones:
Pints, 75 minutes
Quarts, 90 minutes
For jars with bones:
Pints, 65 minutes
Quarts, 75 minutes

POULTRY BROTH

Use chicken, turkey, or any large bird bones. See General Steps in Canning (pp. 19-20).

Hot Pack: Place the large carcass bones in a pot and add enough water to cover bones. Cover pot and simmer for 30 to 45 minutes or until meat can be easily stripped from bones. Remove bones and pick off meat. Cool broth; then, skim off and discard excess fat. Return pieces of meat to broth and heat to boiling. Fill hot jars immediately with meat and broth, leaving a 1-inch headspace. Wipe jar rims and adjust lids.

Process in a pressure canner.
Weighted-gauge canner, 10 pounds pressure
Dial-gauge canner, 11 pounds pressure
Pints, 20 minutes
Quarts, 25 minutes

CLAMS

Keep clams live on ice until ready to can. See General Steps in Canning (pp. 19-20). Scrub shells thoroughly and rinse.

Hot Pack: Steam clams over boiling water for 5 minutes; then, open and remove clam meat. Pour juice from clams into a bowl and save it. Wash clam meat in salt water (1 teaspoon canning salt per quart water) and rinse. Boil enough water in a saucepan to cover clams and add 2 tablespoons lemon juice or ½ teaspoon citric acid per gallon of water. Add clams and boil 2 minutes; then drain. Heat clam juice to boiling. Fill hot jars loosely with clams, leaving a 1-inch headspace. Add boiling clam juice to cover clams, leaving a 1-inch headspace. (Add boiling water to clam juice if there is not enough juice to fill jars.) Remove air bubbles. Wipe jar rims and adjust lids.

Process in a pressure canner.
Weighted-gauge canner, 10 pounds pressure
Dial-gauge canner, 11 pounds pressure
Half-pints, 60 minutes
Pints, 70 minutes

Note: For minced clams, grind them with a meat grinder or food processor after boiling 2 minutes in the lemon water.

CRABS
(Dungeness or King)

Keep live crabs on ice until ready to can. See General Steps in Canning (pp. 19-20). Wash crabs thoroughly, using several changes of cold water.

Hot Pack: In a large pot, prepare a solution of ¼ cup lemon juice and 2 tablespoons canning salt (or up to 1 cup salt, if desired) per gallon water. Add crabs and heat to simmering; simmer for 20 minutes. Place crabs in cold water to cool; then, drain them and remove the back shell. Remove the meat from the body and claws. Prepare a solution of cold water containing 2 tablespoons canning salt (or up to 1 cup salt, if desired) and 2 cups lemon juice per gallon of water. Or, use 4 cups white vinegar (5% acidity) per gallon of water. Soak the meat in the solution for 2 minutes; then, drain and squeeze meat to remove excess moisture. Boil enough water to cover crabs in jars. Fill hot jars with meat, leaving a 1-inch headspace. Add acid to jars: ½ teaspoon citric acid or 2 tablespoons lemon juice to each half-pint; 1 teaspoon citric acid or 4 tablespoons lemon juice per pint. Add boiling water to jars to cover crabs, leaving a 1-inch headspace. Remove air bubbles. Wipe jar rims and adjust lids.

Process in a pressure canner.
Weighted-gauge canner, 10 pounds pressure
Dial-gauge canner, 11 pounds pressure
Half-pints, 70 minutes
Pints, 80 minutes

Note: For best quality, blue crabs should be frozen, not canned.

FISH
(All but Tuna)

Remove internal organs from fish within 2 hours after they are caught. Keep fish on ice until ready to can. See General Steps in Canning (pp. 19-20). Remove heads, tails, fins, and scales. Wash fish and remove all blood. Split fish lengthwise, if desired. Cut cleaned fish into 3½-inch lengths.

Raw Pack: Fill hot jars with raw fish, skin side next to glass, leaving a 1-inch headspace. Add 1 teaspoon canning salt to each pint jar, if desired. Do not add liquid. Wipe jar rims and adjust lids.

Process in a pressure canner.
Weighted-gauge canner, 10 pounds pressure
Dial-gauge canner, 11 pounds pressure
Pints, 100 minutes

Note: Glass-like crystals of magnesium ammonium phosphate sometimes form in canned salmon. There is no way to prevent these crystals from forming when canning at home, but they usually dissolve when heated and are safe to eat.

SMOKED FISH

Use only whitefish, salmon, grayling, and trout. See General Steps in Canning (pp. 19-20). Cut smoked fish into 3½-inch pieces.

Cold Pack: Fill hot jars with fish, leaving a 1-inch headspace. Do not add liquid. Wipe jar rims and adjust lids.

Process in a pressure canner.
Weighted-gauge canner, 10 pounds pressure
Dial-gauge canner, 11 pounds pressure
Pints, 100 minutes

Note: Safe processing times for other smoked fish and seafood have not been determined. Freeze them, instead.

OYSTERS

Keep live oysters on ice until ready to can. See General Steps in Canning (pp. 19-20). Wash shells.

Hot Pack: Heat oysters for 5 to 7 minutes in a preheated oven at 400°F. Cool them briefly in a large container of ice water. Drain oysters, open shells, and remove meat. Wash meat in salt water (½ cup salt per gallon of water); drain. Heat to boiling enough water to cover oysters in jars. Fill hot jars with meat, leaving a 1-inch headspace. Add canning salt to jars, if desired: ¼ teaspoon to half-pints or ½ teaspoon to pints. Add boiling water to cover oysters, leaving a 1-inch headspace. Remove air bubbles. Wipe jar rims and adjust lids.

Process in a pressure canner.
Weighted-gauge canner, 10 pounds pressure
Dial-gauge canner, 11 pounds pressure
Half-pints and Pints, 75 minutes

CANNING NUTS

GREEN PEANUTS IN SHELLS

Use fully mature peanuts that have no insect or disease damage. See General Steps in Canning (pp. 19-20). Wash shells thoroughly with a mild detergent; rinse well in clear water. If peanuts are dry, soak them overnight in a pan of water before canning.

Hot Pack: Boil enough water in a large pot to cover peanuts. Add peanuts and boil 10 minutes. Prepare salt water (8 ounces canning salt to each gallon water) and heat to a boil. Fill hot jars immediately with peanuts, leaving a ½-inch headspace. Add salt water to cover peanuts, leaving a ½-inch headspace. Wipe jar rims and adjust lids.

Process in a pressure canner.
Weighted-gauge canner, 10 pounds pressure
Dial-gauge canner, 11 pounds pressure
Pints, 45 minutes

CHESTNUTS

Use fully mature chestnuts that have no insect or disease damage. See General Steps in Canning (pp. 19-20). Wash shells thoroughly with a mild detergent; rinse well in clear water.

Hot Pack: Cut a gash in the flat side of each shell. Boil enough water in a large pot to cover nuts. Place chestnuts in boiling water and boil for 15 minutes. Drain and cool chestnuts just enough to handle. Remove the shells and brown skins while chestnuts are still hot. Or, heat clean nuts on a baking sheet in a medium-hot oven (350°F) for 15 minutes. Remove shells and skins with a sharp knife. Heat a pot of fresh water to boiling. Fill hot jars immediately with nuts, leaving a ½-inch headspace. Add boiling water to cover chestnuts, leaving a ½-inch headspace. Wipe jar rims and adjust lids.

Process in a boiling water-bath canner.
Pints or Quarts, 30 minutes

Making Pickles and Other Pickled Products

Pickles are made not only from cucumbers, which are most common, but also from many vegetables and fruits, usually mixed with sugar, vinegar, and spices. Some are preserved by fermenting and storing in the refrigerator for several months. Some are processed in a water-bath canner. There are four main types of pickles.

Fermented Pickles. These cucumber pickles are placed in a brine (saltwater) solution to cure for several weeks. They may be stored in the brine solution or they may be soaked in water and then stored in vinegar with spices and sugar added.

Fresh-Pack Pickles. These pickles are made by a quick process. The cucumbers are brined for only several hours or overnight and then drained and combined with boiling vinegar, spices, and other seasonings. Some of the quick-process recipes eliminate brining altogether.

Fruit Pickles. Most pickled fruits are left whole or sliced and simmered in a spicy, sweet-sour syrup made with vinegar or lemon juice.

Relishes. Relishes are made from chopped fruits and vegetables cooked in a spicy vinegar solution.

Caution: The acidity level in pickles or a pickled product is important for safety as well as for taste and texture. Acid prevents the growth of *Clostridium botulinum* bacteria which can cause botulism (see p. 7). Use only tested recipes and do not change the proportions of the vinegar, food, or water.

INGREDIENTS

Vegetables and Fruits

Use tender vegetables and firm fruits. Always use a pickling variety of cucumber. Never try to pickle waxed cucumbers from the supermarket; the brine and pickling solution can't penetrate the wax. Cucumbers should be a uniform size, about 1½ inches long for gherkins and 4 inches for dills. For oddly shaped and larger cucumbers, use recipes that require slicing or cutting up.

Remove a 1/16-inch slice from the blossom end of all vegetables to avoid soft pickles. The blossoms contain enzymes that can cause softening. Do not use any produce that shows the slightest evidence of mold. Mold will cause an off-flavor even though proper processing will kill the potential for spoilage. Wash all produce well, especially around the stems where soil is easily trapped.

For highest quality, pickle fruits or vegetables within 24 hours after they have been picked. Otherwise, refrigerate them to slow the deterioration process.

Salt

Salt is important in the fermentation process. Canning or pickling salt (pure granulated salt) is recommended for making pickles. Table salt is safe to use, but the non-caking materials added to table salt may make the brine cloudy. Flake salt varies in density and is not recommended for use. Reduced-sodium salts may be used in quick pickle recipes; they will be safe to eat, but their quality may be noticeably lower. Both texture and flavor may be different from what you expect. In fermented and brined foods, however, the correct amount of salt is necessary for safety as well as flavor and texture.

Caution: Using less salt than called for or using reduced-sodium salt in fermented pickle or sauerkraut recipes is not recommended. You will not get the proper fermentation without the correct proportion of salt to the other ingredients.

Vinegar

Cider vinegar, the red vinegar, has good flavor and aroma and is used in most pickle recipes. However, it will darken light-colored fruits and vegetables. Distilled vinegar is white or clear. Use white vinegar only when the recipe calls for it. There must be a minimum, uniform level of acid throughout the pickled product to prevent the growth of *Clostridium botulinum* bacteria. Vinegar should have 5-percent acidity or 50-grain strength; this information is on the vinegar label. Never use vinegar that has less than 5-percent acid for making pickles and relishes. They will not be good, and they may spoil. Do not use homemade vinegar and do not dilute vinegar unless the recipe calls for it. If you want to make a recipe less sour, add more sugar rather than less vinegar.

Sugar

Sugar helps to plump pickles and keep them firm. Use white granulated sugar unless the recipe calls for brown sugar. As with cider vinegar, brown sugar adds good flavor but will darken light-colored fruits and vegetables. Using syrup or honey in place of sugar, unless called for in a reliable recipe, may produce an undesirable flavor. Use sugar substitutes **only if** the recipe calls for it. The flavor of commercial sugar substitutes are affected by heat and extended storage.

Spices

Always use fresh spices for best flavor. Tie whole spices in a thin cloth bag before adding them to other ingredients. Then, remove them before pickles and relishes are packed in jars. If spices are packed in the jars, they not only will darken the pickles but also may cause some off-flavor. If you want to substitute fresh spices for dried or dried for fresh: ¼ teaspoon dried spice = 1 teaspoon fresh. For dill: 1 head dill weed = 1 teaspoon dill seed.

Water

Use soft water in the salt-water (brine) solution when making pickles. Hard water can prevent pickles from curing properly; it can interfere with the formation of acid. If you live in a hard-water area, boil the water for 15 minutes and let it sit, covered, for 24 hours. Then, remove any scum that appears and carefully pour the clear water into the brining container to avoid disturbing the sediment. Discard the sediment.

Firming Agents

Though up-to-date methods and good quality ingredients can produce good, crisp pickles without firming agents, you may find some recipes that still call for them. There are two types of firming agents that can be used to make pickles crisp.

Alum. You can buy alum at most supermarkets or drugstores, and it is safe to use when measured correctly. But, alum does not improve the firmness of pickles more than the instructions in the recipes in this book, so it is not included.

Builder's Lime (calcium hydroxide). This is sometimes called slaked lime and can be bought at a hardware or feed store. The calcium in lime can improve pickle firmness. Food-grade lime may be used as a lime-water solution for soaking fresh cucumbers for 12 to 24 hours before pickling them. **Do not use quick lime for pickles.** If too much lime is absorbed by the cucumbers, it must be removed or the pickles will be unsafe to eat. Drain the lime-water solution from the cucumbers, rinse them, and then resoak them in fresh water for 1 hour. Repeat the rinse and soak steps two more times.

Pickles made with lime or alum must be brined in a glass jar or an enamel pot with no chips or cracks in the enamel. Do not put builder's lime or alum in an aluminum boiler or plastic bowl.

METHODS AND PROCEDURES

Boiling Water-Bath Processing

All pickles and pickled products are subject to spoilage from microorganisms, particularly yeasts and molds, as well as enzymes that may affect flavor, color, and texture. Processing in a boiling water-bath canner will prevent both of these problems. Standard canning jars with self-sealing lids are also recommended. Follow the steps in Using the Water-Bath Canner (p. 32). Also, read General Steps in Canning (pp. 19-20) for more complete directions.

Low-Temperature Pasteurization Treatment

Another method for processing cucumber pickles is the low-temperature pasteurization treatment. It can make a crisp pickle with a fairly simple process. However, the correct time and temperature must be maintained to avoid spoilage.

1. Fill hot, sterile jars (see Jars, pp. 21-25) with room-temperature cucumbers.
2. Add hot (165° to 180°F), not boiling, liquid to cover cucumbers, leaving the headspace specified in the recipe.
3. Remove air bubbles, wipe jar rims, and adjust lids (see Lids, pp. 25-26).
4. Place jars on a rack in a large pot or water-bath canner filled half full with warm (120° to 140°F), not boiling, water. Then, add hot water to cover jars by 1 inch.
5. Heat water to maintain a temperature of 180° to 185°F for 30 minutes. Check temperature with a candy or jelly thermometer to be certain water does not fall below 180° or exceed 185°F during the entire 30 minutes.

Caution: Use this method only when suggested in a recipe.

Fermentation

A 1-gallon container is needed for each 5 pounds of fresh vegetables. Therefore, a 5-gallon stone crock is the ideal size for fermenting about 25 pounds of fresh cabbage or cucumbers. Food-grade plastic or glass containers are excellent substitutes for stone crocks. Other 1- to 3-gallon nonfood-grade plastic containers may be used if lined with a clean food-grade plastic bag. **Do not use garbage bags or trash liners.** Sauerkraut can be fermented in quart and half-gallon mason jars, but you may have more spoilage.

1. Wash the container and the other items used in fermentation in hot, sudsy water and rinse them well with very hot water before each filling.
2. Fill the container to within 4 or 5 inches of rim with vegetables and brine, as specified in your recipe.
3. Keep cabbage or cucumbers 1 to 2 inches under the surface of the brine while fermenting. To do this, place a dinner plate or glass pie plate inside the container. The plate should be slightly smaller than the container opening, yet large enough to cover most of the vegetables. Then, place two or three quart jars, filled with water and closed securely with lids, on the plate to weight it down. Or, you can weight the plate down with a very large, clean, food-grade plastic bag filled with 3 quarts water and 2¼ tablespoons salt. Be sure to fasten the bag's seal securely. Using salt water is a precaution, in case the bag leaks accidentally. Freezer bags sold for packaging turkeys are good for use as weights in 5-gallon containers.

4. Cover the container opening with a clean, heavy towel to prevent insects and molds from getting in the jar while the vegetables are fermenting.

5. Place the filled container where temperature is between 70° and 75°F. Temperatures of 55° to 65°F are acceptable, but the fermentation will take 5 to 6 weeks. Avoid temperatures above 80°F because pickles will become too soft during fermentation. Fermenting pickles cure slowly.

6. Check the container several times a week and remove surface scum or mold as soon as it appears.

7. Near the end of the fermentation time, test for bubbles by tapping the container on the side with your hand. Fermentation is complete when bubbles stop coming to the top of the container. As a second test, cut a cucumber in half crosswise; if it is the same color throughout and has no noticeable rings, fermentation is complete.

Caution: If the pickles become soft, slimy, or develop a disagreeable odor, discard them.

MATERIALS AND EQUIPMENT

For Preparing:
- Stainless steel, glass, or enamel saucepan
- Large wooden, plastic, or stainless steel spoons
- Food chopper or grinder
- Colander or sieve
- Large-mouth funnel
- Scales
- Large measuring cup
- Sharp knife
- Clean cloths and towels
- Hot pads

For Canning:
- Water-bath canner
- Standard canning jars
- Two-piece lids
- Jar lifter
- Lid wand
- Jar funnel
- Food-grade plastic bags

For Fermenting:
- Stoneware crock, glass jar, or food-grade container
- Heavy plate or glass lid
- Jars with lids for weights

Caution: Not all stoneware crocks are heat-sealed, especially decorative crocks that have come on the market in recent years. Liquid will seep through the crock if it is not heat-sealed, and it is not safe to use.

COMMON CAUSES OF PROBLEMS

Soft or Slippery Pickles

Pickles may be soft or slippery if:
- the brine or vinegar is too weak
- the vegetable is exposed above the brine while curing
- the fermentation container is stored at a temperature above 75°F during curing
- the vegetable is cooked too long or at too high a temperature
- the blossom ends of vegetables are not removed
- the garlic or spices are moldy
- the pickles are not processed long enough or at the correct temperature

Shriveled Pickles

Shriveling often occurs when:
- the cucumbers went through drought-stress during development
- the cucumbers are held longer than 24 hours between picking and brining
- the cucumbers are placed in a very strong salt, sugar, or vinegar solution without going through the proper stages
- the cucumbers are cooked or processed too fast or too long

Hollow Pickles

Hollow pickles may be the result when:
- the cucumbers did not develop correctly
- the cucumbers are held longer than 24 hours between picking and brining
- the cucumbers are too large for curing
- the brine is not kept at the proper strength
- the cucumbers are exposed above the brine while curing
- the curing process is ended before fermentation is complete

Spoiled Vegetables in Brine

DO NOT EAT ANY PICKLE THAT SHOWS SIGNS OF SPOILAGE! The top layer of vegetables fermenting in brine can spoil if:
- the scum which forms on the surface of the brine is not removed frequently. Scum is made up of yeasts, molds, and bacteria which will attack and break down the structure of the vegetables beneath. Scum may also weaken the acidity of the brine, which also causes spoilage

Cloudy, Dark, or Discolored Pickles

These conditions in the appearance of pickles usually result when:
- poor quality cucumbers are used

- brass, iron, copper, or zinc pots or utensils are used—DO NOT EAT PICKLES THAT WERE PREPARED IN OR WITH BRASS, COPPER, OR ZINC UTENSILS
- the water is hard, containing iron or other minerals
- table salt is used
- the brine is not kept at the proper strength
- ground spices are used or whole spices are packed in the jars with pickles
- the curing process is ended before fermentation is complete
- pickles are stored in a place that is too warm or too light

Bitter Taste

A strong or bitter taste can occur when:
- the cucumbers went through drought-stress during development
- the vinegar is too acid
- the spices are cooked too long in the vinegar
- there are too many spices
- a salt substitute containing potassium chloride is used

PICKLE RECIPES

SAUERKRAUT
Makes about 9 quarts.

25 pounds cabbage ¾ cup pickling salt

Use firm, mature heads of cabbage. Plan to get cabbage ready to place in the fermentation container between 24 and 48 hours after it is harvested. Work with about 5 pounds of cabbage at a time. Remove and discard the outer leaves. Rinse heads under cold running water and drain. Cut heads into quarters and remove cores. Shred or slice cabbage about ¼ inch thick. Put cabbage in a fermentation container (see Fermentation, pp. 97-98). Add 3 tablespoons salt. Mix salt with cabbage thoroughly with your hands. Pack cabbage firmly until salt draws juices from cabbage. Repeat shredding, salting, and packing until all cabbage is in the container. Be sure container is deep enough to allow 4 or 5 inches between the cabbage and the rim. If juice does not cover cabbage, add boiled and cooled brine (1½ tablespoons pickling salt per quart of water). Add plate and weight it. Cover container with a clean towel. Store at 70° to 75°F for about 3 to 4 weeks. If you weight the cabbage with a brine-filled bag, do not disturb the container until normal fermentation is complete. If you use jars as weight, check the kraut 2 to 3 times each week and remove scum if it forms. Fully fermented kraut may be kept tightly covered in the refrigerator for several months. Or, it may be canned.

To Can Sauerkraut: See General Steps in Canning (pp. 19-20).

Hot Pack: Pour kraut and liquid into a large pot. Bring it to a boil over medium heat, stirring frequently. Fill hot jars rather firmly with kraut, leaving a ½-inch headspace. Add boiling liquid to cover kraut, leaving a ½-inch headspace. Remove air bubbles. Wipe jar rims and adjust lids.

Process in a boiling water-bath canner.
Pints, 10 minutes
Quarts, 15 minutes

Raw Pack: Drain and boil liquid in a saucepan. Fill hot jars firmly with kraut, leaving a ½-inch headspace. Add boiling liquid to cover kraut, leaving a ½-inch headspace. Remove air bubbles. Wipe jar rims and adjust lids.

Process in a boiling water-bath canner.
Pints, 10 minutes
Quarts, 25 minutes

DILL PICKLES

4 pounds pickling cucumbers
2 tablespoons dill seed
2 garlic cloves (optional)
2 dried red peppers (optional)
2 teaspoons whole mixed pickling spice (optional)
½ cup pickling salt
¼ cup vinegar (5% acidity)
8 cups water

Use 4-inch cucumbers. Wash them well. Cut a $1/16$-inch slice off blossom ends and discard. Leave a ¼-inch stem attached. Place half of the dill seed, garlic, peppers, and pickling spice in the bottom of a fermentation container (see Fermentation, pp. 97-98). Add cucumbers, remaining dill, garlic, pepper, and spice. Dissolve salt in vinegar and water and pour over cucumbers. Add plate to cover cucumbers and weight it. When fermentation is complete, pickles may be stored in the fermentation container for up to 6 months in the refrigerator, if you are careful to remove surface scum and molds on a regular basis. However, canning fermented pickles is a better way to store them.

To Can Pickles: See General Steps in Canning (pp. 19-20).

Raw Pack: Pour the brine into a large saucepan. Heat slowly to a boil and simmer 5 minutes. To reduce cloudiness, pour the brine through paper coffee filters. Fill hot jars with pickles, leaving a ½-inch headspace. Add boiling brine to cover pickles, leaving a ½-inch headspace. Remove air bubbles. Wipe jar rims and adjust lids.

Process in a boiling water-bath canner.
Pints, 10 minutes
Quarts, 15 minutes

Note: Quantities listed in recipe are for a 1-gallon fermentation container.

REFRIGERATOR DILLS
Makes about 5 quarts.

6 pounds pickling cucumbers
18 to 24 large heads of fresh dill weed
1½ gallons water
¾ cup pickling salt
2 to 3 garlic cloves, peeled and sliced
6 tablespoons mixed pickling spice

Use 4-inch cucumbers. Wash them well. Cut a 1/16-inch slice from blossom ends and discard. Leave a ¼-inch stem attached. Place cucumbers in a 3-gallon fermentation container (see Fermentation, pp. 97-98). Add dill. In a large pot, combine water, salt, garlic, and pickling spices. Bring to a boil. Cool and pour over cucumbers in container. Cover with a plate and weight it. Keep it at 70° to 75°F for 1 week. Then, fill jars with pickles and cover with pickling liquid. Close jars securely with lids and store in the refrigerator. Pickles may be eaten after 3 days and should be used within 2 months.

QUICK FRESH-PACK DILL PICKLES
Makes about 7 quarts.

18 pounds pickling cucumbers
½ cup pickling salt
2 gallons water
1½ quarts vinegar (5% acidity)
¾ cup pickling salt
¼ cup sugar
2 quarts water
2 tablespoons whole mixed pickling spice
5 tablespoons whole mustard seed
21 heads of fresh dill

Use 3- to 5-inch cucumbers. See General Steps in Canning (pp. 19-20). Wash cucumbers. Cut a 1/16-inch slice off the blossom ends and discard. Leave a ¼-inch stem attached. Place cucumbers in a large pot. Dissolve ½ cup salt in 2 gallons water. Pour salt water over cucumbers and let them soak 12 hours. Drain.

Raw Pack: Combine vinegar, ¾ cup salt, sugar, and 2 quarts water in a large pot. Tie mixed pickling spice in a clean white cloth and add to pot. Heat to boiling to make a pickling syrup. Fill hot jars with cucumbers, leaving a ½-inch headspace. Add 1 teaspoon mustard seed and 1½ heads fresh dill to each pint jar; 2 teaspoons mustard seed and 3 heads dill to each quart. Add boiling syrup to cover cucumbers, leaving a ½-inch headspace. Remove air bubbles. Wipe jar rims and adjust lids.

Process in a boiling water-bath canner.
Pints, 10 minutes
Quarts, 15 minutes

For low-temperature pasteurization treatment, see page 97.

REDUCED-SODIUM SLICED DILL PICKLES
Makes about 8 pints.

4 pounds pickling cucumbers	2 tablespoons pickling salt
2 large onions	1½ teaspoons celery seed
6 cups cider vinegar (5% acidity)	1½ teaspoons mustard seed
6 cups sugar	8 heads fresh dill weed

Use 3- to 5-inch cucumbers. See General Steps in Canning (pp. 19-20). Wash cucumbers. Cut a 1/16-inch slice off blossom ends and discard. Cut cucumbers into ¼-inch slices. Cut onions into thin slices.

Raw Pack: Combine vinegar, sugar, salt, celery seed, and mustard seed in a large pot. Bring to a boil to make a pickling syrup. Place 2 onion slices and ½ dill head in each hot jar. Fill jars with cucumbers, leaving a ½ -inch headspace. Add 1 onion slice and ½ dill head on top. Add boiling syrup to cover cucumbers, leaving a ¼-inch headspace. Remove air bubbles. Wipe jar rims and adjust lids.

Process in a boiling water-bath canner.
Pints, 15 minutes

BREAD AND BUTTER PICKLES
Makes about 8 pints.

6 pounds pickling cucumbers	4½ cups sugar
3 pounds onions	2 tablespoons mustard seed
½ cup pickling salt	1½ tablespoons celery seed
Crushed ice or ice cubes	1 tablespoon ground turmeric
4 cups cider vinegar (5% acidity)	1 cup pickling lime (optional; see below)

Use 4- to 5-inch cucumbers. See General Steps in Canning (pp. 19-20). Wash cucumbers. Cut a 1/16-inch slice off the blossom ends and discard. Cut cucumbers into 3/16-inch slices. Peel onions and cut them into thin slices. Combine cucumbers and onions in a large bowl. Add salt. Cover with 2 inches of ice. Refrigerate for 3 to 4 hours, adding more ice as needed.

Hot Pack: Combine vinegar, sugar, mustard seed, celery seed, and turmeric in a large pot. Heat to boiling and boil 10 minutes to make a pickling syrup. Add drained cucumbers and onions to boiling syrup. Return to a boil over medium heat. Fill hot jars with cucumbers and onions, leaving a ½-inch headspace. Add boiling syrup to cover cucumbers, leaving a ½-inch headspace. Remove air bubbles. Wipe jar rims and adjust lids.

Process in a boiling water-bath canner.
Pints and Quarts, 10 minutes
For low-temperature pasteurization treatment, see page 97.

Note: Store jars for 4 to 5 weeks before using to develop characteristic flavor.

To Use Pickling Lime: Follow directions above, but after cutting cucumbers into ³⁄16-inch slices, mix 1 cup pickling lime with ½ cup salt and 1 gallon water in a 2- to 3-gallon crock or enamelware container. Soak cucumber slices in the lime water for 12 to 24 hours, stirring occasionally. Then, remove cucumbers from lime solution, rinse, and soak for 1 hour in fresh cold water. Repeat fresh-water rinsing and soaking procedures two more times. Handle slices carefully because they will be brittle. Drain well. Cover onions with ice and refrigerate for 3 to 4 hours, adding more ice as needed. Then, follow directions above for the Hot Pack method.

Caution: Avoid inhaling lime dust while mixing the lime-water solution.

To Vary: Use slender (1- to 1½-inch diameter) zucchini or yellow summer squash in place of cucumbers.

14-DAY SWEET PICKLES
(Slices, Strips, or Whole)
Makes about 5 pints whole cucumbers; 9 pints slices.

4 pounds pickling cucumbers
Water
¾ cup pickling salt
2 teaspoons celery seed
2 tablespoons mixed pickling spice
5½ cups sugar
4 cups vinegar (5% acidity)

Use 2- to 5-inch cucumbers. See General Steps in Canning (pp. 19-20). Wash cucumbers. Cut a ¹⁄16-inch slice off blossom ends and discard. Leave a ¼-inch stem attached if making whole pickles. Place whole cucumbers in a 1-gallon fermentation container (see Fermentation, pp. 97-98). In a saucepan, combine 2 quarts water and ¼ cup of the salt and bring to a boil. Pour salt water over cucumbers. Cover cucumbers and weight them down. Place a clean towel over the container and keep the temperature at about 70°F.

On the third and fifth days, drain the salt water and discard. Rinse cucumbers and scald the container, the cover, and weight. Return cucumbers to container. Boil another 2 quarts water with ¼ cup salt. Pour over cucumbers. Replace cover and weight; cover container with a clean towel.

On the seventh day, drain salt water again and discard. Rinse cucumbers and scald container, cover, and weight. If desired, slice cucumbers or cut them in strips. Then, return them to the container. Tie celery seed and pickling spice loosely in a spice bag. Combine 2 cups of the sugar and 4 cups vinegar in a saucepan to make a syrup. Add spice bag, bring to a boil, and pour syrup over cucumbers. Add cover and weight; cover container with a clean towel.

On each of the next six days, drain syrup and spice bag into a saucepan. Add ½ cup of the remaining sugar each day and bring to a boil. Remove cucumbers and rinse. Scald container, cover, and weight daily. Return cucumbers to container, add hot syrup, cover, and weight; cover with a clean towel.

Making Pickles • 105

On the 14th day, prepare canner and jars. Drain syrup into saucepan. Add ½ cup sugar to syrup and bring to a boil. Fill hot jars with cucumbers, leaving a ½-inch headspace. Remove spice bag from syrup. Add boiling syrup to cover cucumbers, leaving a ½-inch headspace. Remove air bubbles. Wipe jar rims and adjust lids.

Process in a boiling water-bath canner.
Pints, 5 minutes
Quarts, 10 minutes
For low-temperature pasteurization treatment, see page 97.

QUICK SWEET PICKLES
(Slices or Strips)
Makes about 8 pints.

8 pounds pickling cucumbers	2 teaspoons celery seed
⅓ cup pickling salt	1 tablespoon whole allspice
Crushed ice or ice cubes	2 tablespoons mustard seed
4½ cups sugar	1 cup pickling lime (optional; see below
3½ cups vinegar (5% acidity)	and Caution on page 106)

Use 3- to 4-inch cucumbers. See General Steps in Canning (pp. 19-20). Wash cucumbers. Cut a ¹⁄₁₆-inch slice off both ends and discard. Slice cucumbers or cut them into strips and place in a bowl. Sprinkle with salt. Cover with 2 inches of ice. Refrigerate 3 to 4 hours, adding more ice as needed. Drain well. Combine sugar, vinegar, celery seed, allspice, and mustard seed in a large pot. Heat to boiling and boil 10 minutes to make a pickling syrup.

Hot Pack: Add cucumbers to boiling syrup and heat to a boil on medium heat. Stir occasionally to be sure mixture heats evenly. Fill hot jars with cucumbers, leaving a ½-inch headspace. Add boiling syrup to cover cucumbers, leaving a ½-inch headspace. Remove air bubbles. Wipe jar rims and adjust lids.

Process in a boiling water-bath canner.
Pints and Quarts, 5 minutes

Raw Pack: Fill hot jars with cucumbers, leaving a ½-inch headspace. Add boiling syrup, leaving a ½-inch headspace. Remove air bubbles. Wipe jar rims and adjust lids.

Process in a boiling water-bath canner.
Pints, 10 minutes
Quarts, 15 minutes
For low-temperature pasteurization treatment, see page 97.

To Vary: Add 2 slices of raw onion to each jar before filling with cucumbers.

To Use Pickling Lime: Follow directions above but instead of placing cucumber slices or strips in a bowl covered with ice, soak them in a mixture of 1 cup pickling lime and ½ cup pickling salt to each gallon of water in a 2- to 3-gallon container for 12 to 24 hours, stirring occasionally. Then, remove cucumbers from lime solution, rinse, and soak 1 hour in fresh, cold water. Repeat the fresh-water rinsing and soaking procedures two more times. Handle cucumbers carefully because they will be brittle. Drain well. Then, follow directions on previous page for making pickling syrup and processing.

Caution: Avoid inhaling lime dust while mixing the lime solution.

SWEET GHERKIN PICKLES
Makes about 7 pints.

7 pounds pickling cucumbers	2 teaspoons celery seed
Water	2 teaspoons whole mixed pickling spice
½ cup pickling salt	2 cinnamon sticks
8 cups sugar	½ teaspoon fennel (optional)
6 cups vinegar (5% acidity)	2 teaspoons vanilla (optional)
¾ teaspoon turmeric	

This process will take four days to complete. It is best to begin in the morning of the first day. Use 1½-inch, or smaller, cucumbers. See General Steps in Canning (pp. 19-20). Wash cucumbers. Cut a ¹⁄₁₆-inch slice from the blossom ends and discard. Leave a ¼-inch stem attached. Place cucumbers in a fermentation container (see Fermentation, pp. 97-98) and add enough boiling water to cover them. **After 6 to 8 hours,** drain cucumbers, discard water, and scald container. Then, place cucumbers back in container. Combine ¼ cup of the salt with 6 quarts fresh boiling water and again cover cucumbers. Let them soak about 24 hours.

About 24 hours later (second day), drain cucumbers again and discard brine. Scald container. Place cucumbers in container and cover with another ¼ cup salt in 6 quarts of fresh boiling water.

The morning of third day, drain cucumbers and discard brine. Scald container. In a large pot, combine 3 cups of the sugar, 3 cups of the vinegar, the turmeric, celery seed, pickling spice, cinnamon sticks, and fennel. Heat to a boil to make a pickling syrup. Prick each cucumber with a table fork and place them in the container. Pour boiling syrup over cucumbers. This mixture will not completely cover the cucumbers. **After 6 to 8 hours,** drain the syrup into a saucepan and add 2 cups more sugar and 2 cups more vinegar. Heat to a boil and pour over cucumbers in the container.

The morning of the fourth day, drain syrup again into a saucepan. Add another 2 cups sugar and 1 cup vinegar. Heat to a boil and pour over cucumbers. After 6 to 8 more hours, prepare canner and jars. Drain syrup into saucepan again and add the last cup of sugar and the vanilla. Heat to boiling. Fill hot jars with cucumbers, leaving a ½-inch headspace. Add boiling syrup to cover cucumbers, leaving a ½-inch headspace. Remove air bubbles. Wipe jar rims and adjust lids.

Process in a boiling water-bath canner.
Pints, 5 minutes
For low-temperature pasteurization treatment, see page 97.

REDUCED-SODIUM SLICED SWEET PICKLES
Makes about 4 pints.

4 pounds pickling cucumbers
1⅔ cups white vinegar (5% acidity)
3 cups sugar
1 tablespoon whole allspice
2¼ teaspoons celery seed

1 quart white vinegar (5% acidity)
1 tablespoon pickling salt
1 tablespoon mustard seed
½ cup sugar

Use 3- to 4-inch cucumbers. See General Steps in Canning (pp. 19-20). Wash cucumbers. Cut a ¹⁄₁₆-inch slice off blossom ends and discard. Cut cucumbers into ¼-inch slices.

Hot Pack: Combine 1⅔ cups vinegar, 3 cups sugar, allspice, and celery seed in a saucepan to make a pickling syrup. Bring to a boil and keep syrup hot until used. In a large pot, mix 1 quart vinegar, salt, mustard seed, and ½ cup sugar. Add the cucumber slices, cover, and simmer until they change color from bright to dull green, about 5 to 7 minutes. Drain cucumbers and discard cooking liquid. Fill hot jars with cucumbers, leaving a ½-inch headspace. Add boiling syrup to cover cucumbers, leaving a ½-inch headspace. Remove air bubbles. Wipe jar rims and adjust lids.

Process in a boiling water-bath canner.
Pints, 10 minutes

PICKLE RELISH
Makes about 9 pints.

3 quarts chopped cucumbers
3 cups chopped green bell peppers
3 cups chopped sweet red peppers
1 cup chopped onions
¾ cup pickling salt
Crushed ice or ice cubes
Water

4 teaspoons mustard seed
4 teaspoons turmeric
4 teaspoons whole allspice
4 teaspoons whole cloves
2 cups sugar
6 cups white vinegar (5% acidity)

See General Steps in Canning (pp. 19-20). Combine cucumbers, red and green peppers, onions, salt, 4 cups ice, and 2 quarts water in a large pot. Let them soak for 4 hours; then, drain. Cover vegetables with the same amount of fresh ice and water and let them soak for another hour. Drain again. Tie mustard seed, turmeric, allspice, and cloves loosely in a spice bag.

Hot Pack: Combine sugar and vinegar in a saucepan. Add spice bag and heat to boiling to make a pickling syrup. Stir to dissolve sugar. Pour hot syrup over vegetables. Cover and refrigerate 24 hours. Then, heat vegetable mixture and syrup to boiling. Fill hot jars with relish, leaving a ½-inch headspace. Remove air bubbles. Wipe jar rims and adjust lids.

Process in a boiling water-bath canner.
Half-pints and Pints, 10 minutes

PICKLED PEPPER ONION RELISH
Makes about 9 half-pints.

3 cups finely chopped sweet red peppers
3 cups finely chopped green bell peppers
6 cups finely chopped onions
1½ cups sugar
6 cups white vinegar (5% acidity)
2 tablespoons pickling salt

See General Steps in Canning (pp. 19-20).

Hot Pack: Combine all ingredients in a large pot and boil gently about 30 minutes or until mixture thickens and volume is reduced by one-half. Fill hot jars with relish, leaving a ½-inch headspace. Remove air bubbles. Wipe jar rims and adjust lids.

Process in a boiling water-bath canner.
Half-pints, 5 minutes

Note: Jars of relish that will be used within one month do not need to be processed. Simply cover them securely with lids and store in the refrigerator.

PICCALILLI
Makes about 3 pints.

6 cups chopped green tomatoes
1½ cups chopped sweet red peppers
1½ cups chopped green peppers
2¼ cups chopped onion
7½ cups chopped cabbage
½ cup pickling salt
Water
3 tablespoons whole mixed pickling spice
4½ cups cider vinegar (5% acidity)
3 cups brown sugar

See General Steps in Canning (pp. 19-20). In a large bowl, combine tomatoes, peppers, onion, cabbage, and salt. Add enough hot water to cover vegetables and let them soak 12 hours. Drain and press vegetables in a clean white cloth to remove all possible liquid. Tie pickling spice loosely in a spice bag.

Hot Pack: Combine vinegar and brown sugar in a large pot. Add spice bag and heat to a boil. Add vegetables and boil gently 30 minutes or until the volume of the mixture is reduced by one-half. Remove spice bag. Fill hot jars with relish, leaving a ½-inch headspace. Remove air bubbles. Wipe jar rims and adjust lids.

Process in a boiling water-bath canner.
Pints, 5 minutes

PICKLED GREEN TOMATO RELISH
Makes about 8 pints.

10 pounds green tomatoes
1½ pounds sweet red peppers
1½ pounds green bell peppers
2 pounds onions
½ cup pickling salt

1 quart water
4 cups sugar
1 quart cider vinegar (5% acidity)
⅓ cup prepared mustard
2 tablespoons cornstarch

Use small, hard tomatoes. See General Steps in Canning (pp. 19-20). Wash and coarsely grate or finely chop tomatoes, red and green peppers, and onions.

Hot Pack: Place grated vegetables in a large pot. Dissolve salt in water and pour over vegetables. Heat to boiling, reduce heat, and simmer 5 minutes. Drain through a colander or sieve. Return vegetables to pot. Add sugar, vinegar, mustard, and cornstarch. Stir to mix. Heat to boiling, reduce heat, and simmer 5 minutes. Fill hot jars with relish, leaving a ½-inch headspace. Remove air bubbles. Wipe jar rims and adjust lids.

Process in a boiling water-bath canner.
Pints, 5 minutes

CORN RELISH
Makes about 9 pints.

16 to 20 ears of corn
2½ cups diced sweet red peppers
2½ cups diced green bell peppers
2½ cups chopped celery
1¼ cups chopped onion
1¾ cups sugar

5 cups white vinegar (5% acidity)
2½ tablespoons pickling salt
2½ teaspoons celery seed
2½ tablespoons dry mustard
1¼ teaspoons turmeric

Use medium-size ears of corn. See General Steps in Canning (pp. 19-20). Remove husks, wash corn, and remove silks.

Hot Pack: Place corn in a large pot of water and boil 5 minutes. Then, dip the ears in cold water and cut kernels from cob. Prepare enough to make 10 cups of kernels. Combine red and green peppers, celery, onion, sugar, vinegar, salt, and celery seed in a large pot. Bring to a boil, reduce heat, and simmer 5 minutes, stirring occasionally. Dip out ½ cup of the simmered mixture and add mustard and turmeric. Mix well. Then, return this mixture to the pot and add corn. Simmer another 5 minutes. If desired, thicken mixture with flour paste (¼ cup flour blended in ¼ cup water) and stir frequently. Fill hot jars with relish, leaving a ½-inch headspace. Remove air bubbles. Wipe jar rims and adjust lids.

Process in a boiling water-bath canner.
Half-pints or Pints, 15 minutes

Note: You can use six 10-ounce packages of frozen corn in place of fresh corn.

DIXIE RELISH
Makes about 5 pints.

4 large sweet red peppers	2 quarts water
4 large green bell peppers	¾ cup sugar
1 small head cabbage	2 tablespoons mustard seed
2 large white onions	2 tablespoons celery seed
¼ cup pickling salt	1 quart cider vinegar (5% acidity)

See General Steps in Canning (pp. 19-20). Wash red and green peppers; cut them in half, remove seeds and cores, and chop them. Prepare enough to make 4 cups of mixed peppers. Remove outside leaves of cabbage and chop it in small pieces. Peel and wash onions and chop enough to make 2 cups. Mix chopped vegetables in a large glass or enamel pot. Dissolve salt in water and pour over vegetables. Let them soak for about 1 hour; then, drain thoroughly.

Hot Pack: Combine sugar, mustard seed, celery seed, and vinegar in a large pot. Heat to boiling. Add drained vegetables and simmer 20 minutes. Fill hot jars with relish, leaving a ½-inch headspace. Remove air bubbles. Wipe jar rims and adjust lids.

Process in a boiling water-bath canner.

Pints, 15 minutes

GARDEN RELISH
Makes about 4 pints.

1 quart chopped cabbage	3 tablespoons pickling salt
3 cups chopped cauliflower	3¾ cups cider vinegar (5% acidity)
2 cups chopped green tomatoes	2¾ cups sugar
2 cups chopped onion	3 teaspoons celery seed
2 cups chopped green bell pepper	3 teaspoons dry mustard
1 cup chopped sweet red pepper	1½ teaspoons turmeric

See General Steps in Canning (pp. 19-20). Combine chopped vegetables in a large bowl or flat baking pan. Sprinkle the salt over vegetables and let them sit for 4 to 6 hours in a cool place. Then, drain off accumulated juices. Combine vinegar, sugar, celery seed, mustard, and turmeric in a large pot. Heat slowly and simmer 10 minutes. Add vegetables and simmer 10 more minutes. Then, increase heat and quickly bring to a boil. Fill hot jars with relish, leaving a ½-inch headspace. Remove air bubbles. Wipe jar rims and adjust lids.

Process in a boiling water-bath canner.

Pints, 10 minutes

ONION RELISH
Makes about 3 pints.

3 pounds onions
Water
1 cup white vinegar (5% acidity)
1 cup sugar
1 teaspoon pickling salt
½ teaspoon mustard seed

See General Steps in Canning (pp. 19-20). Peel and slice enough onions to make 8 cups.

Hot Pack: Boil enough water in a saucepan to cover onions. Drop slices in boiling water and boil 4 minutes. Drain. In a large saucepan, mix vinegar, sugar, salt, and mustard and heat to a boil. Add onion slices and simmer 4 minutes. Fill hot jars with onion slices, leaving a ½-inch headspace. Add boiling liquid to cover onions, leaving a ½-inch headspace. Remove air bubbles. Wipe jar rims and adjust lids.

Process in a boiling water-bath canner.
Pints, 10 minutes

PICKLED HORSERADISH SAUCE
Makes about 2 half-pints.

¾ pound horseradish roots
1 cup white vinegar (5% acidity)
½ teaspoon pickling salt
¼ teaspoon powdered ascorbic acid

Wash roots thoroughly and peel off brown outer skin. Grate peeled roots in a food processor. Or, cut them into small cubes and put them through a grinder. Prepare enough to make 2 cups grated horseradish. Combine grated horseradish with vinegar, salt, and ascorbic acid. Fill hot jars, leaving a ¼-inch headspace. Close jars securely with lids and store in the refrigerator.

Note: Make only small quantities of this sauce at a time because the pungency of fresh horseradish fades within 1 to 2 months, even when refrigerated.

PEAR RELISH
Makes about 10 pints.

1 peck pears
6 large onions
6 green bell peppers
6 sweet red peppers
1 bunch celery
3 cups sugar
1 tablespoon allspice
1 tablespoon salt
5 cups vinegar

See General Steps in Canning (pp. 19-20). Wash pears; peel and cut them in half. Remove cores. Wash and peel onions. Wash green and red peppers; cut them in half and remove stems, seeds, and cores. Separate and clean the celery stalks. Put pears, onions, peppers, and celery through a food chopper. Combine the sugar, allspice, salt, and vinegar in a large pot. Add the chopped pear mixture and mix well. Let it sit overnight.

Hot Pack: Heat mixture to a boil. Fill hot jars with boiling relish, leaving a ½-inch headspace. Remove air bubbles. Wipe jar rims and adjust lids.
Process in a boiling water-bath canner.
Pints, 20 minutes

PICKLED DILLED BEANS
Makes about 8 pints.

4 pounds green or yellow beans
4 cups white vinegar (5% acidity)
4 cups water
½ cup pickling salt

1 teaspoon hot red pepper flakes (optional)
8 to 16 heads fresh dill
8 garlic cloves (optional)

Use fresh beans, 5 to 6 inches long. See General Steps in Canning (pp. 19-20). Wash beans and trim ends. Cut beans to 4-inch lengths.

Raw Pack: In a large saucepan, combine vinegar, water, salt, and pepper flakes. Bring to a boil to make a pickling liquid. Place 1 to 2 dill heads and 1 garlic clove in each jar. Place whole beans upright in jars, leaving a ½-inch headspace. Trim beans to ensure proper headspace, if necessary. Add boiling liquid to beans, leaving a ½-inch headspace. Remove air bubbles. Wipe jar rims and adjust lids.
Process in a boiling water-bath canner.
Pints, 5 minutes

PICKLED THREE-BEAN SALAD
Makes about 3 pints.

1 pound green or yellow beans
1½ cups canned, drained red kidney beans
½ cup white vinegar (5% acidity)
¼ cup bottled lemon juice
¾ cup sugar
1¼ cups water

¼ cup olive oil or salad oil
½ teaspoon pickling salt
1 cup canned, drained garbanzo beans
½ cup thinly sliced onion
½ cup thinly sliced celery
½ cup sliced green bell pepper

Use fresh beans. See General Steps in Canning (pp. 19-20). Wash beans and cut off ends. Cut or snap them into 1- to 2-inch pieces. Prepare enough to make 1½ cups.

Hot Pack: Boil enough water in a large pot to just cover beans. Add beans and blanch 3 minutes. Remove beans from pot immediately to cool. Rinse kidney beans with tap water and drain again. In a large pot, combine vinegar, lemon juice, sugar, and 1¼ cups water; heat to a boil. Remove from heat, add oil and salt, and mix well. Add all the beans, onion, celery, and green pepper and heat to simmering. Then, remove pot from heat and let vegetables marinate in the refrigerator for 12 to 14 hours. Heat bean mixture again to a boil. Fill hot jars with the vegetables, leaving a ½-inch headspace. Add boiling liquid to cover vegetables, leaving a ½-inch headspace. Remove air bubbles. Wipe jar rims and adjust lids.

Process in a boiling water-bath canner.
Half-pints and Pints, 15 minutes

PICKLED BEETS
Makes about 8 pints.

7 pounds beets	2 cups sugar
4 to 6 onions (optional)	2 cups water
4 cups cider vinegar (5% acidity)	2 cinnamon sticks, crushed
1½ teaspoons pickling salt	12 whole cloves

Use 2- to 2½-inch diameter beets and onions. See General Steps in Canning (pp. 19-20). Trim off beet tops, leaving a 1-inch stem, and leave the roots to prevent color bleeding. Wash beets thoroughly.

Hot Pack: Boil enough water in a large pot to cover beets. Add beets and cook until tender, about 25 to 30 minutes. Drain beets and discard the cooking liquid. Cool beets and trim off roots and stems. Slip off skins. Slice beets into ¼-inch slices. Peel and thinly slice onions. Combine vinegar, salt, sugar, and water in a large pot. Tie cinnamon and cloves in a spice bag and drop in vinegar mixture. Bring to a boil and add beets and onions. Simmer 5 minutes. Remove spice bag. Fill hot jars with beets and onions, leaving a ½-inch headspace. Add boiling liquid to cover beets, leaving a ½-inch headspace. Remove air bubbles. Wipe jar rims and adjust lids.

Process in a boiling water-bath canner.
Pints or Quarts, 30 minutes

To Vary: Follow directions for pickled beets but use 1 to 1¼-inch diameter whole beets. Pack beets whole; do not slice.

PICKLED CAULIFLOWER
Makes about 9 half-pints.

12 cups cauliflower flowerets	1 teaspoon hot red pepper flakes
4 cups white vinegar (5% acidity)	1 teaspoon turmeric
2 cups sugar	2 tablespoons mustard seed
2 cups thinly sliced onion	1 tablespoon celery seed
1 cup diced sweet red pepper	

Use flowerets that are 1 to 2 inches across the top. See General Steps in Canning (pp. 19-20). Wash cauliflower.

Hot Pack: Place cauliflower in a large pot. Cover with salt water (4 teaspoons pickling salt per gallon of water) and heat to a boil. Boil 3 minutes; then, drain and discard cooking liquid. Cool cauliflower. In a large saucepan, combine vinegar, sugar, onion, red pepper, pepper flakes, turmeric, mustard seed, and celery seed. Bring to a boil, reduce heat, and simmer 5 minutes to make a pickling syrup. Place an even mix of onion and red pepper in hot jars. Then, fill jars with cauliflower, leaving a ½-inch headspace. Add boiling syrup to cover cauliflower, leaving a ½-inch headspace. Remove air bubbles. Wipe jar rims and adjust lids.

Process in a boiling water-bath canner.
Half-pints or Pints, 10 minutes

To Vary: Use small brussels sprouts in place of cauliflower. Remove stems and blemished outer leaves. Follow directions for cauliflower except boil sprouts in salt water 4 minutes instead of 3.

MARINATED WHOLE MUSHROOMS
Makes about 9 half-pints.

7 pounds small whole mushrooms
½ cup bottled lemon juice
2 cups olive oil or salad oil
2½ cups white vinegar (5% acidity)
1 tablespoon oregano leaves
1 tablespoon dried basil leaves

1 tablespoon pickling salt
½ cup finely chopped onion
¼ cup diced pimiento
2 garlic cloves
25 black peppercorns

Use very fresh unopened mushrooms with caps less than 1¼ inches in diameter. See General Steps in Canning (pp. 19-20). Wash mushrooms and trim stems, leaving only ¼ inch attached to cap.

Hot Pack: Place mushrooms in a large pot. Add lemon juice and enough water to cover. Bring to a boil, reduce heat, and simmer 5 minutes. Drain mushrooms and discard cooking liquid. Mix oil, vinegar, oregano, basil, and salt in a saucepan. Stir in onion and pimiento and heat to boiling. Cut garlic into quarters and place one quarter garlic and 2 or 3 peppercorns in each hot jar. Fill jars with mushrooms, leaving a ½-inch headspace. Add boiling, well-mixed liquid to cover mushrooms, leaving a ½-inch headspace. Remove air bubbles. Wipe jar rims and adjust lids.

Process in a boiling water-bath canner.
Half-pints, 20 minutes

PICKLED DILLED OKRA
Makes about 9 pints.

7 pounds small okra pods
⅔ cup pickling salt
6 small hot peppers
4 teaspoons dill seed

6 cups water
6 cups white vinegar (5% acidity)
8 to 9 garlic cloves

See General Steps in Canning (pp. 19-20). Wash and trim okra.

Raw Pack: Combine salt, hot peppers, dill seed, water, and vinegar in a large pan and bring to a boil. Fill hot jars firmly with whole okra, leaving a ½-inch headspace. Put 1 garlic clove in each hot jar. Add boiling liquid to cover okra, leaving a ½-inch headspace. Remove air bubbles; wipe jar rims, adjust lids.

Process in a boiling water-bath canner. Pints, 10 minutes

PICKLED HOT PEPPERS
See Note and Caution, below. Makes about 9 pints.

4 pounds long hot peppers
1½ pounds sweet red peppers
1½ pounds green bell peppers
5 cups white vinegar (5% acidity)

1 cup water
4 teaspoons pickling salt
2 tablespoons sugar
2 garlic cloves

See General Steps in Canning (pp. 19-20). Wash hot peppers. Small ones may be left whole; large ones may be quartered. Cut two to four slits in each pepper to allow steam to escape.

Hot Pack: To remove the tough skin on hot peppers, blister them using the Oven or Range-Top Method.

Oven or Broiler Method: Place hot peppers on a baking sheet or flat pan in a hot oven (400°F) or under a broiler for 6 to 8 minutes until skins blister.

Range-Top Method: Cover hot burner, either gas or electric, with heavy wire mesh. Place hot peppers on burner for several minutes until skins blister.

Allow hot peppers to cool; then, place them in a pan and cover with a damp cloth. This will make peeling the peppers easier. After several minutes, peel each pepper. Discard seeds; flatten whole peppers. Wash sweet red and green bell peppers. Cut them in half and remove seeds and cores. Boil enough water in a large saucepan to cover sweet peppers. Add peppers and blanch 3 minutes. Drain. Combine vinegar, water, salt, sugar, and garlic in a large saucepan. Heat to boiling, reduce heat, and simmer 10 minutes to make a pickling liquid. Fill hot jars with evenly mixed peppers, leaving a ½-inch headspace. Remove garlic from liquid. Add boiling liquid to cover peppers, leaving a ½-inch headspace. Remove air bubbles. Wipe jar rims and adjust lids.

Process in a boiling water-bath canner. Half-pints and Pints, 10 minutes

Note: Includes banana, chili, Hungarian, and jalapeno peppers.

Caution: Wear rubber gloves while handling hot peppers and wash your hands thoroughly with soap and water before touching your face.

MARINATED HOT PEPPERS
See Note and Caution, on page 116. Makes about 9 half-pints.

4 pounds peppers
1 cup olive oil or salad oil
2 cups white vinegar (5% acidity)
1 cup bottled lemon juice
1 tablespoon oregano leaves

½ cup chopped onion
2 garlic cloves, quartered (optional)
2 tablespoons prepared horseradish (optional)
Pickling salt

Use firm peppers with no blemishes. See General Steps in Canning (pp. 19-20). Wash peppers. Small ones may be left whole; large ones may be quartered. Cut two to four slits in each whole pepper to allow steam to escape.

Hot Pack: To remove the tough skin on hot peppers, blister them using the Oven or Range-Top Method.

Oven or Broiler Method: Place peppers on a baking sheet or flat pan in a hot oven (400°F) or under a broiler for 6 to 8 minutes until skins blister.

Range-Top Method: Cover hot burner, either gas or electric, with heavy wire mesh. Place peppers on burner for several minutes until skins blister.

Allow peppers to cool; place them in a pan, cover with a damp cloth. This will make peeling easier. After several minutes, peel each pepper. Discard seeds; flatten whole peppers. In a saucepan, mix the oil, vinegar, lemon juice, oregano, onion, garlic, and horseradish; heat to boiling to make a pickling liquid. Place a garlic quarter in each hot jar. Add pickling salt, if desired: ¼ teaspoon per half-pint or ½ teaspoon per pint. Fill hot jars with peppers, leaving a ½-inch headspace. Mix boiling liquid well and add to cover peppers, leaving a   -inch headspace. Remove air bubbles. Wipe jar rims, adjust lids.

Process in a boiling water-bath canner. Half-pints or Pints, 15 minutes

Note: Includes banana, chili, Hungarian, and jalapeno peppers.

To Vary: The intensity of marinated jalapeno peppers may be adjusted as follows: hot, 4 pounds jalapeno peppers; medium hot, 2 pounds jalapeno peppers and 2 pounds sweet and mild peppers; mild, 1 pound jalapeno peppers and 3 pounds sweet and mild peppers. Sweet or mild peppers may be marinated using this recipe. However, they do not need to have skins removed. Instead, they should be placed in boiling water and blanched 3 minutes.

Caution: Wear rubber gloves while handling hot peppers and wash your hands thoroughly with soap and water before touching your face.

PEPPER SAUCE
Makes about 9 pints

1 gallon small red or green hot peppers	2 tablespoons sugar	1 tablespoon horseradish
1½ cups salt	1 clove garlic	1 cup water
		5 cups vinegar

Wash and drain peppers. Cut two small slits in each pepper. Dissolve salt in 1 gallon water. Pour over peppers. Let stand 12 to 18 hours. Rinse and drain. Add sugar, garlic, horseradish, and 1 cup water to vinegar. Simmer 15 minutes. Remove garlic. Pack peppers into hot standard canning jars. Heat pickling liquid to boiling and pour over peppers. Adjust jar lids and bands. Process in boiling waterbath canner (212°F) for 10 minutes.

PICKLED BELL PEPPERS
Makes about 9 pints.

7 pounds bell peppers
3½ cups sugar
3 cups white vinegar (5% acidity)

3 cups water
9 garlic cloves, halved
Pickling salt

Use firm red or green peppers that have no blemishes. See General Steps in Canning (pp. 19-20). Wash peppers and cut them into quarters. Remove cores and seeds, and cut away any small spots. Slice peppers into strips.

Hot Pack: In a large pot, boil sugar, vinegar, and water for 1 minute to make a pickling syrup. Add peppers and return to a boil. Place one garlic half in each hot jar. Add pickling salt to jars, if desired: ¼ teaspoon per half-pint or ½ teaspoon per pint. Fill hot jars with peppers, leaving a ½-inch headspace. Add boiling syrup to cover peppers, leaving a ½-inch headspace. Remove air bubbles. Wipe jar rims and adjust lids.

Process in a boiling water-bath canner.
Half-pints or Pints, 10 minutes

PICKLED SWEET GREEN TOMATOES
Makes about 9 pints.

10 pounds green tomatoes
2 cups sliced onions
¼ cup pickling salt
4 cups cider vinegar (5% acidity)
3 cups brown sugar

1 tablespoon mustard seed
1 tablespoon allspice
1 tablespoon celery seed
1 tablespoon whole cloves

See General Steps in Canning (pp. 19-20). Wash and slice tomatoes. Prepare enough to make about 16 cups. Mix tomato and onion slices in a bowl and sprinkle with salt. Let them sit for 4 to 6 hours; then, drain thoroughly.

Hot Pack: Heat vinegar in a large pot and add brown sugar to make a pickling syrup. Stir until sugar dissolves. Tie mustard seed, allspice, celery seed, and cloves loosely in a spice bag. Add bag to hot syrup. Add tomatoes and onions. If needed, add just enough water to cover slices. Bring to a boil, reduce heat, and simmer 30 minutes, stirring as needed to prevent burning. Tomatoes should become tender and transparent. Remove spice bag. Fill hot jars with tomatoes and onions, leaving a ½-inch headspace. Add boiling syrup, leaving a ½-inch headspace. Remove air bubbles. Wipe jar rims and adjust lids.

Process in a boiling water-bath canner.
Pints, 10 minutes
Quarts, 15 minutes

PICKLED MIXED VEGETABLES
Makes about 10 pints.

4 pounds pickling cucumbers
2 pounds small onions
4 cups 1-inch celery slices
2 cups ½-inch carrot slices, peeled
2 cups ½-inch sweet red pepper squares
2 cups cauliflower flowerets
Crushed ice or ice cubes
5 cups white vinegar (5% acidity)
¼ cup prepared mustard
½ cup pickling salt
3½ cups sugar
3 tablespoons celery seed
2 tablespoons mustard seed
½ teaspoon whole cloves
½ teaspoon ground turmeric

Use 4- to 5-inch cucumbers. See General Steps in Canning (pp. 19-20). Wash cucumbers. Cut a 1/16-inch slice off both ends and discard. Cut cucumbers into 1-inch slices. Peel and quarter onions. Combine cucumber slices and onions with celery, carrots, red pepper, and cauliflower in a large bowl. Cover with 2 inches of ice and refrigerate for 3 to 4 hours, adding more ice if needed.

Hot Pack: In an large pot, combine vinegar and mustard and mix well. Add salt, sugar, celery seed, mustard seed, cloves, and turmeric. Bring to a boil to make a pickling syrup. Drain vegetables and add to boiling syrup. Cover and slowly return to a boil. Drain vegetables and save the syrup. Fill hot jars with an even mix of vegetables, leaving a ½-inch headspace. Add boiling syrup to cover vegetables, leaving a ½-inch headspace. Remove air bubbles. Wipe jar rims and adjust lids.

Process in a boiling water-bath canner.
Pints, 5 minutes
Quarts, 10 minutes

BREAD AND BUTTER ZUCCHINI PICKLES
Makes about 9 pints.

16 cups sliced fresh zucchini
4 cups thinly sliced onions
½ cup pickling salt
Water
4 cups white vinegar (5% acidity)
2 cups sugar
4 tablespoons mustard seed
2 tablespoons celery seed
2 teaspoons ground turmeric

See General Steps in Canning (pp. 19-20). Combine zucchini and onion slices in a large pot; add salt and enough water to cover vegetables by 1 inch. Let vegetables soak for 2 hours; then, drain thoroughly.

Hot Pack: In a large pot, combine vinegar, sugar, mustard seed, celery seed, and turmeric. Heat to a boil to make a pickling syrup. Add zucchini and onions and simmer 5 minutes. Fill hot jars with an even mix of zucchini and onions, leaving a ½-inch headspace. Add boiling syrup to cover vegetables, leaving a ½-inch headspace. Remove air bubbles. Wipe jar rims and adjust lids.

Process in a boiling water-bath canner.
Pints and Quarts, 10 minutes
For low-temperature pasteurization treatment, see page 97.

SPICED APPLE RINGS
Makes about 9 pints.

12 pounds tart apples	1¼ cups white vinegar (5% acidity)
12 cups sugar	3 tablespoons whole cloves
6 cups water	¾ cup red hot cinnamon candies

Use firm apples that are no larger than 2½ inches in diameter. See General Steps in Canning (pp. 19-20). Wash apples. To help prevent darkening, peel only one apple at a time and immediately slice it horizontally into ½-inch slices. Remove the core area and place the slices into an ascorbic acid solution (see Pretreating, p. 39).

Hot Pack: In a large saucepan, combine the sugar, water, vinegar, cloves, and cinnamon candies. Stir and heat to a boil to make a pickling syrup. Reduce heat and simmer 3 minutes. Drain apple slices, add them to the hot syrup, and cook 5 minutes. Fill hot wide-mouth jars immediately with apple rings, leaving a ½-inch headspace. Add boiling syrup to cover apples, leaving a ½-inch headspace. Remove air bubbles. Wipe jar rims and adjust lids.

Process in a boiling water-bath canner.
Half-pints or Pints, 10 minutes

Note: You can use 8 cinnamon sticks and 1 teaspoon red food coloring in place of the cinnamon candies.

SPICED CRAB APPLES
Makes about 9 pints.

5 pounds crab apples	4 teaspoons whole cloves
4½ cups cider vinegar (5% acidity)	4 cinnamon sticks
3¾ cups water	6 ½-inch cubes fresh gingerroot
7½ cups sugar	

See General Steps in Canning (pp. 19-20). Remove blossom petals but leave stems attached. Wash apples. Puncture the skin of each apple four times with an ice pick or toothpick.

Hot Pack: Mix vinegar, water, and sugar in a large pot. Use a pot that has a blancher basket or will hold a sieve or colander. Bring mixture to a boil to make a pickling syrup. Tie cloves, cinnamon, and gingerroot loosely in a spice bag and add to boiling syrup. Place one-third of the apples at a time into the blancher basket or sieve. Immerse the apples in the boiling syrup for 2 minutes. Then, remove and repeat with the next third. When all apples have been boiled, place them in a clean 1- or 2-gallon crock; add the spice bag and hot syrup. Cover and let them plump overnight. The next day, remove spice bag and drain syrup into a large saucepan. Heat syrup to boiling. Fill hot jars immediately with apples, leaving a ½-inch headspace. Add boiling syrup to cover apples, leaving a ½-inch headspace. Remove air bubbles. Wipe jar rims and adjust lids.

Process in a boiling water-bath canner.
Pints, 20 minutes.

PEACH PICKLES
Makes about 6 pints.

8 pounds peaches
1 quart cider vinegar (5% acidity)
6¾ cups sugar
4 2-inch cinnamon sticks, crushed
2 tablespoons whole cloves
1 tablespoon ground ginger

Use small to medium-size peaches. See General Steps in Canning (pp. 19-20). Wash and peel peaches. To prevent darkening, place them in an ascorbic acid solution (see Pretreating, p. 39).

Hot Pack: Pour vinegar into a large pot. Add sugar and stir to make a pickling syrup. Heat to a boil and boil 5 minutes; skim off foam. Tie cinnamon sticks, cloves, and ginger loosely in a spice bag. Add bag to the syrup. Remove peaches from solution and drain well. Place peaches into syrup and boil until they can be pierced with a fork but are not soft. Remove pan from heat and allow peaches to plump overnight in syrup. Next day, reheat peaches to a boil and remove spice bag. Fill hot jars with peaches, leaving a ½-inch headspace. Add boiling syrup to cover peaches, leaving a ½-inch headspace. Remove air bubbles. Wipe jar rims and adjust lids.

Process in a boiling water-bath canner.
Pints, 20 minutes

PEAR PICKLES
Makes about 7 pints.

8 pounds pears
8 cups sugar
1 quart white vinegar (5% acidity)
2 cups water
8 2-inch cinnamon sticks
2 tablespoons whole cloves
2 tablespoons whole allspice

Use Seckel pears or some other pickling variety. See General Steps in Canning (pp. 19-20). Wash and peel pears. To prevent darkening, prepare only one pear at a time and place it into ascorbic acid solution (see Pretreating, p. 39). Cut off the blossom end of each pear; the stem may be left on, if desired. If the pear is large, cut it in half or quarter it.

Hot Pack: In a large pot, combine sugar, vinegar, water, and cinnamon sticks to make a pickling syrup. Tie cloves and allspice loosely in a spice bag and add bag to the syrup mixture. Heat to a boil, reduce heat, and simmer, covered, about 30 minutes. Remove pears from solution and rinse well. Add pears to syrup and continue simmering for 20 to 25 more minutes. Fill hot jars with pears, leaving a ½-inch headspace. Place a cinnamon stick in each jar. Add boiling syrup to cover pears, leaving a ½-inch headspace. Remove air bubbles. Wipe jar rims and adjust lids.

Process in a boiling water-bath canner.
Pints, 20 minutes

WATERMELON RIND PICKLES
Makes about 4 pints.

3 quarts watermelon rind	3 cups white vinegar (5% acidity)
¾ cup pickling salt	3 cups water
3 quarts water	1 tablespoon whole cloves
2 trays ice cubes	6 1-inch cinnamon sticks
9 cups sugar	1 lemon, thinly sliced

See General Steps in Canning (pp. 19-20). Trim away any remaining peel or pink edges from the rind. Cut white rind into 1-inch squares. Mix salt and 3 quarts water in a large bowl. Add rind and ice cubes and let them sit for 3 to 4 hours. Then, drain and rinse in cold water.

Hot Pack: Place rind in a large pot and cover with cold water. Heat to boiling, reduce heat, and simmer about 10 minutes. Do not overcook. Drain, wash pot, and return rind to pot. Combine sugar, vinegar, and 3 cups water in a saucepan to make a pickling syrup. Tie cloves and cinnamon sticks loosely in a spice bag; add bag to syrup. Heat to a boil and boil 5 minutes. Pour syrup over rind. Remove seeds from lemon and add slices. Let rind plump overnight in syrup. Next day, heat again to boiling, reduce heat, and cook slowly for 1 hour. Remove spice bag. Fill hot jars with watermelon rind, leaving a ½-inch headspace. Place a cinnamon stick in each jar. Add boiling syrup to cover rind, leaving a ½-inch headspace. Remove air bubbles. Wipe jar rims and adjust lids.

Process in a boiling water-bath canner.
Pints, 10 minutes

PICKLED EGGS

1 dozen medium-size eggs	1 teaspoon pickling salt
2 teaspoons mustard	1 teaspoon celery seed
2 cups cider vinegar (5% acidity)	1 teaspoon mustard seed
½ cup water	6 whole cloves
1 cup sugar	2 medium onions, sliced

Hard-cook eggs by boiling them for 8 minutes. Cool and crack the entire shell around the egg for easier peeling. Begin peeling the shell at the large end where the air cell is usually located. Peeling under running water will help remove all the thin shell membrane. Place eggs in a quart jar or any quart container with a tight-fitting lid. In a saucepan, combine the remaining ingredients. Heat the mixture to near boiling and simmer for 5 minutes. Pour the hot mixture over the eggs. Close jar securely with lid and store immediately in the refrigerator.

Note: Pickled eggs may be kept several months in the pickling solution in refrigerator. Drain before serving.

RED BEET EGGS

1 dozen medium-size eggs
1 cup red beet juice
½ cup cider vinegar (5% acidity)
1 teaspoon brown sugar
Several canned whole or sliced red beets

Follow directions for Pickled Eggs (p. 122).

DILLED EGGS

1 dozen medium-size eggs
1½ cups white vinegar (5% acidity)
1 cup water
¾ teaspoon dill seed
¼ teaspoon white pepper
3 teaspoons pickling salt
¼ teaspoon mustard seed
½ teaspoon onion juice
½ teaspoon minced garlic

Follow directions for Pickled Eggs (p. 122).

SWEET AND SOUR EGGS

1 dozen medium-size eggs
1½ cups apple cider
½ cup cider vinegar (5% acidity)
1 12-ounce package red cinnamon candy
1 tablespoon mixed pickling spice
2 teaspoons pickling salt
1 teaspoon garlic salt

Follow directions for Pickled Eggs (p. 122).

SPICY EGGS

1 dozen medium-size eggs
1½ cups apple cider
1 cup white vinegar (5% acidity)
2 teaspoons pickling salt
1 teaspoon mixed pickling spice
1 garlic clove, peeled
½ medium onion, sliced
½ teaspoon mustard seed

Follow directions for Pickled Eggs (p. 122).

DARK AND SPICY EGGS

1 dozen medium-size eggs
1½ cups cider vinegar (5% acidity)
½ cup water
1 tablespoon dark brown sugar
2 teaspoons granulated sugar
1 teaspoon mixed pickling spice
¼ teaspoon liquid smoke or hickory smoke salt
2 teaspoons pickling salt

Follow directions for Pickled Eggs (p. 122).

ROSY PICKLED EGGS

1 dozen medium-size eggs
1½ cups white vinegar (5% acidity)
½ cup water
½ teaspoon pickling salt
6 whole cloves
1 bay leaf
1 onion, sliced
8 to 10 drops red food coloring
1 cup sugar

Follow directions for Pickled Eggs (p. 122).

Making Jelly and Other Jellied Spreads

Jellied spreads have many textures, flavors, and colors, depending on the kind of fruit used and the way it is prepared. Sugar is the primary preserving agent. Jellied spreads made from crushed or ground fruit are described below.

Jelly is a clear mixture of fruit juice and sugar. It is semi-solid but firm enough to hold its shape.

Jam is made from a crushed or chopped fruit and sugar. It is not clear and contains small pieces of the fruit throughout. Jam will hold its shape, but it is less firm than jelly.

Conserves are like jams but are made from a mixture of fruits, especially citrus fruits, with nuts, raisins, or coconut.

Preserves are made of small, whole fruits or uniform-size pieces in a clear, thick, slightly jellied syrup.

Marmalades are soft fruit jellies with small pieces of fruit or citrus peel suspended in the clear jelly.

Fruit butters are made from fruit pulp that has been cooked with sugar until thickened to a spreadable consistency.

INGREDIENTS

For proper texture, jellied fruit spreads require the correct combination of fruit, pectin, acid, and sugar.

Fruit

The fruit gives each spread its unique flavor and color. It also supplies the liquid needed to dissolve the rest of the necessary ingredients and furnishes some or all of the pectin and acid. Good-quality, flavorful fruits make the best jellied products.

Pectin

Pectins are substances in fruits that form a gel if they are mixed in the right combination with acid and sugar. All fruits contain some pectin. Sour apples, crab apples, sour blackberries, gooseberries, loganberries, cranberries, lemons, quinces, currants, and some plums and grapes usually contain enough natural pectin to form a gel if they are not overripe. All other fruits and berries contain little pectin; they must be combined with fruits high in pectin or with commercial pectin products to gel. As fruits ripen, they lose their pectin. Therefore, for jellies that will be made with no commercial pectin, one-fourth of the fruits should be underripe.

Acid

Acidity is also critical to gel formation. If there is too little acid, the gel will never set; if there is too much acid, the gel will lose its liquid (weep). The fruits listed above which have enough natural pectin also have enough natural acid to gel. For fruits low in acid, add lemon juice or other acid ingredients as directed in specific recipes. Commercial pectin products contain acids to help ensure gelling.

Sugar

Sugar serves as a preserving agent, contributes flavor, and aids gelling. Cane and beet sugars are the usual ingredients in jellied spreads. Corn syrup and honey may be used to replace part of the sugar in recipes, but too much will mask the fruit flavor and alter the gel structure. Use tested recipes for replacing sugar with honey or corn syrup. Do not try to reduce the amount of sugar in traditional recipes. Too little sugar prevents gelling and may allow yeasts and molds to grow.

Using Less Sugar. Jellied spreads that contain modified pectin, gelatin, or gums may be made with noncaloric sweeteners. Jams made with concentrated fruit pulp may be made with less sugar than usual because the pulp contains less liquid and less sugar.

Modified Pectin. Two types of modified pectin are available for home use. One gels with one-third less sugar. The other is a low-methoxyl pectin which requires a source of calcium for gelling. To prevent spoilage, jars with these products must be processed longer in a boiling water-bath canner. Recipes and processing times provided with each modified pectin product must be followed carefully. The proportions of acids and fruits should not be altered because spoilage may result. The flavor, texture, and appearance of the jellied product made with low-methoxyl pectin will not be the same as traditional jellies and jams.

Refrigerated Fruit Spreads. Some recipes have been developed for jellied fruit spreads (p. 162) made with gelatin and sugar substitutes. Such products spoil at room temperature and, therefore, must be kept refrigerated. They should be eaten within one month.

GENERAL STEPS IN MAKING JELLY

1. Selecting Fruit. Select a mixture of about three-fourths ripe and one-fourth slightly underripe or firm-ripe fruit. Slightly underripe fruit contains more acid and pectin than ripe, but ripe fruit furnishes more of the desired flavor. Do not use overripe fruit at all—the juice may fail to gel because of its low acid and pectin content, and the flavor may not be good. One pound of prepared fruit makes about 1 cup of juice. One cup of juice, made with an equal amount (1 cup) of sugar, makes about 1½ cups of jelly.

Do not use commercially canned or frozen fruit juices. Their pectin content is too low.

2. Preparing Containers. The best containers for jelly are half-pint standard canning jars with 2-piece lids (see Jars, pp. 21-25, and Lids, pp. 25-26). Wash jars in warm, soapy water and rinse well. Sterilize jars (p. 22) because jelly is processed in a boiling water-bath canner (see Using the Water-Bath Canner, p. 32) for less than 10 minutes, not enough time to kill all bacteria on the jars. Keep jars hot. You can do this by placing them in the canner when you are preheating the water for processing.

3. Preparing Fruit. Wash fruits thoroughly and discard all damaged parts. Wash berries quickly and carefully. Leave skins on grapes and plums. Remove stems and blossom ends of apples and quince. Cut fruit into pieces, but do not remove cores or skins; they add pectin to the juice when they are cooking.

Prepare small quantities of fruit at a time. For example, cook at one time about 4 quarts of berries or 4 pounds of apples or grapes. If you plan to make up a larger quantity of jelly, start cooking a second batch of fruit as soon as the first has finished dripping from the jelly bag.

Table 12. Extracting Juice: Cups of Water Per Pound of Fruit*

Fruit	Cups of Water
Hard Fruits Apples, Crab Apples, Pears, Quince	1 or enough to cover fruit
Soft Fruits Plums, Some Peaches	½
Wild Grapes	1
Cultivated Grapes such as Concord	¼ to None
Firm Berries	¼
Soft Berries	None

*One pound of fruit should yield about 1 cup of clear juice.

4. Extracting Juice. Fruit must be cooked to extract the pectin. In cooking juicy fruit, add only the amount of water specified in your recipe (or see Table 12, above). If too much water is used, the excess has to be cooked out. Prolonged cooking destroys pectin, flavor, and color.

If the fruit lacks its normal juiciness, as may occur when it is grown under drought conditions, double the proportion of water and increase cooking time to soften the fruit.

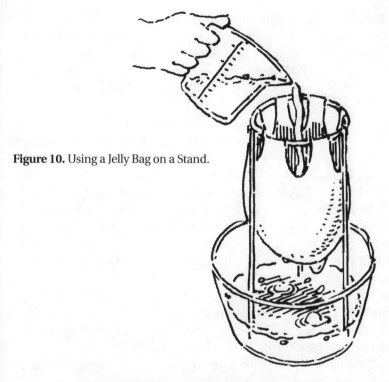

Figure 10. Using a Jelly Bag on a Stand.

Heat fruit in a large, heavy-bottom saucepan and stir to prevent scorching. Crush soft fruits or berries to start the flow of juice. When fruit begins to boil, reduce heat and simmer until it is soft, about 10 minutes for berries and grapes, 20 to 25 minutes for harder fruits.

When fruit is tender, strain it through a colander. Then, strain it through a jelly bag or a double layer of cheesecloth. Use a stand or colander to support the bag. Let juice drip through bag without pressing it. Pressing or squeezing the bag will cause jelly to be cloudy.

Note: Small tartaric acid (tartrate) crystals will form in grape juice. Let juice stand in the refrigerator overnight. Strain out crystals before canning.

5. Testing Juice. If you are not following an exact recipe, you may want to test for pectin and acid after extracting juice. If the fruit you are using does not have enough of either, you will need to add them. Otherwise, your jelly will not gel properly, and all your efforts will be wasted. There are three ways to test for pectin and one test for acid.

Cooking Test. Measure ⅓ cup juice and ¼ cup sugar into a small saucepan. Heat slowly, stirring constantly until the sugar dissolves. Increase the heat and boil rapidly until it reaches the sheeting stage (see Sheet or Spoon Test, pp. 128-129). Pour the jelly into a small bowl and let it cool. If the cooled mixture is jelly-like, there is enough pectin and the juice will gel.

Alcohol Test. Pour 1 tablespoon rubbing alcohol into a jar with a tight-fitting lid. Add 1 teaspoon of the juice. Close the jar and gently shake the mixture so that all the juice comes in contact with the alcohol. **DO NOT TASTE**—this mixture is poisonous. If the juice forms into a solid jelly-like mass that can be picked up with a fork, it has enough pectin. If the juice clumps into several small particles, it does not have enough pectin to gel.

Jelmeter Test. The jelmeter is a glass tube that measures the rate at which the juice flows through the tube. It can give a rough estimate of the amount of pectin in the juice and how much sugar should be used. Your Extension county agent can tell you how to get a jelmeter.

Testing for Acid. Mix 1 teaspoon lemon juice, 3 tablespoons water, and ½ teaspoon sugar and taste it. Then, taste your fruit juice. If the juice is not as tart as the lemon mixture, it is not tart enough, and acid needs to be added when making jelly.

Purchase fresh pectin each year. Old pectin may result in a poor gel. Follow the instructions with each package.

6. Preparing Canner. The water-bath canner is recommended for making jellied fruit spreads. Preheat the canner by filling it half full with water (see Using The Water-Bath Canner, p. 32) and heating it. Heat another pot of water to boiling to add to the canner after adding jars.

7. Cooking Jelly. Using no more than 6 to 8 cups of juice at a time, measure the ingredients in the recipe you are using. Heat to boiling over high heat. Stir to help dissolve sugar. Then, boil rapidly (on high heat) until the jelly is done or has reached the jellying point.

8. Testing for Doneness. Use one of the following methods to test jelly for doneness.

Temperature Test. Use a jelly or candy thermometer and boil until mixture reaches 220°F. If the altitude where you are working is above 2,000 feet, you

Table 13. Jellying Point at Critical Altitude Differences

Sea Level	2,000 Feet	4,000 Feet	6,000 Feet	8,000 Feet
220°F	216°F	212°F	209°F	205°F

will need to adjust for altitude (see Table 13, above). No place in Alabama has an altitude above 1,000 feet.

Sheet or Spoon Test. Using a cool metal spoon, dip out a spoonful of the boiling jelly mixture. Raise the spoon about 12 inches above the pan, out of the steam. Turn the spoon slightly so the liquid begins to run off the side. The jelly is done if the syrup forms two drops that flow together and "sheet" or hang off the edge of the spoon.

Making Jelly • 129

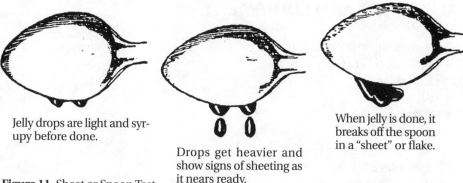

Jelly drops are light and syrupy before done.

Drops get heavier and show signs of sheeting as it nears ready.

When jelly is done, it breaks off the spoon in a "sheet" or flake.

Figure 11. Sheet or Spoon Test

Refrigerator-Freezer Test. Pour a small amount of the boiling jelly on a small plate. Put it in the freezing unit of your refrigerator for a few minutes. If it gels, it should be done. While you are doing this test, remove the pan of jelly mixture from the heat.

9. Canning Jelly. Remove mixture from heat and quickly skim off foam. Fill hot jars immediately with jelly, leaving a ¼-inch headspace (see Checking Headspace, pp. 31-32). Use a measuring cup to fill jars or ladle the jelly through a wide-mouth funnel. Wipe jar rims and adjust lids. Process in a boiling water-bath canner in half-pints or pints for 5 minutes.

Never double a recipe. If a double batch of jelly is processed for only 5 minutes, it will be undercooked. If processed longer, it will turn dark and have a caramelized flavor.

Important Points to Remember

1. There are two distinct jobs to jelly making: extracting the juice and cooking the juice with sugar to make jelly.

2. If a clear jelly is desired, don't squeeze the jelly bag when extracting the juice. Allow the juice to drip from it.

3. Have all necessary utensils on hand before beginning to cook jelly. See that all jelly glasses are clean, sterilized, and hot before the jelly is ready to be poured into them.

4. You get better results when only a few cups of juice are boiled at one time. Boil jelly rapidly; don't simmer it.

5. More jelly is spoiled from the use of too much sugar than from too little. Measure accurately.

6. Just as soon as a good jelly test is reached, remove jelly from heat, skim, and pour into jars immediately.

MATERIALS AND EQUIPMENT
Needed:
- Large, thick-bottom saucepan
- Measuring cup
- Colander
- Sieve
- Jelly bag
- Scales
- Standard canning jars
- Two-piece lids
- Water-bath canner
- Timer or clock
- Clean cloths

Helpful:
- Support for jelly bag
- Jar lifter
- Jar funnel
- Lid wand
- Quart measuring cup
- Food blender or food mill
- Potato masher
- Clean towels
- Hot pads

COMMON CAUSES OF PROBLEMS

Formation of Crystals
Sugar or tartrate (in grape juice) crystals form in jelly because:
- grape juice is not allowed to sit overnight
- too much sugar is used
- undissolved sugar sticks to the side of the pan while cooking
- mixture is cooked for too long or over heat that is too low
- not enough acid is in fruit
- the jelly is left open too long before sealing

Syneresis or "Weeping"
Weeping can take place when:
- too much acid is in the juice
- the jelly is stored in a warm place

Mold
DO NOT EAT JELLIED SPREADS THAT HAVE MOLDS GROWING ON THEM! Molds may grow on jellied spreads if:
- the jar is not sterile when filled
- the jar is not properly sealed

Fermentation
DO NOT EAT JELLIED SPREADS THAT HAVE FERMENTED! Yeasts or bacteria cause fermentation of jellied spreads when:
- the jelly is not properly processed in a boiling water-bath canner
- the jar is not properly sealed

Bubbles
DO NOT EAT JELLIED SPREADS IN WHICH BUBBLES ARE MOVING! Bubbles can be the result when:
- the pot is held too high when pouring jelly into the jar
- the jelly is poured too slowly into the jar
- the jar is not properly sealed

Soft or Syrupy Jelly
A soft or syrupy jelly may be the result if:
- too large a batch of fruit is cooked at one time
- the fruit is overcooked when extracting juice
- too much water is used to extract juice
- too much or too little sugar is added to the juice
- not enough pectin or acid is in juice
- the juice is not cooked to the jellying point

Stiff or Tough Jelly
If jelly is stiff or tough, it may be because:
- too much underripe fruit is used, resulting in too much pectin
- too little sugar is used for the pectin in the juice
- the jelly is cooked beyond the jellying point

Tough Fruit in Preserves
The fruit pieces in jellied fruit spreads can be tough when:
- the fruit is placed in too heavy a syrup to start cooking
- the fruit is not plumped long enough
- the fruit is overcooked

Shriveled Fruit in Preserves
The fruit pieces in jellied fruit spreads can be shriveled when:
- the syrup is too heavy (too much sugar) for the fruit used.

Gummy Jelly

Gummy jelly can be the result if:
- the jelly is simmered rather than boiled rapidly
- the jelly is overcooked

Strong or Bitter Flavor

Strong or bitter flavors occur if:
- the fruit is not good quality
- too large a batch of fruit is cooked at one time
- the fruit is overcooked when extracting juice
- the fruit is not stirred enough to prevent scorching

Cloudy Jelly

Cloudy jelly is often the result when:
- the fruit is too underripe and contains starch
- the juice is not strained well enough
- the foam is not skimmed off the jelly
- the jelly is not poured up immediately and partially congeals

Dark or Discolored Jelly

Jellied fruit spreads can appear dark or discolored if:
- too large a batch of fruit is cooked at one time
- the fruit is cooked too slowly
- the fruit is overcooked
- the jellied spread is not stored properly

JELLY RECIPES

APPLE JELLY
Makes 4 or 5 half-pints.

3 pounds apples
3 cups water
2 tablespoons lemon juice (optional)
3 cups sugar

Use about one-fourth slightly underripe and three-fourths ripe apples. See General Steps in Making Jelly (pp. 126-129). Wash apples and remove stem and blossom ends; do not peel or core them. Cut apples into small pieces and remove seeds. Place apple pieces in a large, thick-bottom saucepan.

Add water, cover, and bring to a boil on high heat. Reduce heat and simmer 20 to 25 minutes or until apples are soft. Extract juice (pp. 126-127). Measure 4 cups apple juice into a saucepan. Add lemon juice and sugar and stir well. Bring to a boil on high heat. Boil rapidly until jelly is done (see Testing For Doneness, pp. 128-129). Remove from heat and quickly skim off foam. Fill hot jars immediately with jelly, leaving a ¼-inch headspace. Wipe jar rims and adjust lids.

Process in a boiling water-bath canner.
Half-pints, 5 minutes

BLACKBERRY JELLY
Makes 4 or 5 half-pints.

2½ quarts blackberries   3 cups sugar
¾ cup water

Use about one-fourth slightly underripe and three-fourths fully ripe berries. See General Steps in Making Jelly (pp. 126-129). Wash berries and discard bad ones. Remove any stems and caps. Crush berries in a saucepan. Add water, cover, and bring to a boil on high heat. Reduce heat and simmer 5 minutes. Extract juice (pp. 126-127). Measure 4 cups berry juice into a saucepan. Add sugar and stir well. Bring to a boil on high heat. Boil rapidly until jelly is done (see Testing for Doneness, pp. 128-129). Remove from heat and quickly skim off foam. Fill hot jars immediately with jelly, leaving a ¼-inch headspace. Wipe jar rims and adjust lids.

Process in a boiling water-bath canner.
Half-pints, 5 minutes

BLUEBERRY JELLY
Makes 7 or 8 half-pints.

2 quarts blueberries   12 ounces liquid pectin
7½ cups sugar

See General Steps in Making Jelly (pp. 126-129). Wash blueberries and discard bad ones. Thoroughly crush berries, one layer at a time, in a large, thick-bottom saucepan. Bring to a boil on high heat; cover and simmer 5 minutes, stirring occasionally. Extract juice (pp. 126-127). Measure 4 cups berry juice into a large saucepan. Stir in sugar. Place on high heat, stirring constantly, and bring to a full rolling boil that cannot be stirred down. Add the liquid pectin and heat again to a full rolling boil. Boil rapidly for 1 minute. Remove from heat and quickly skim off foam. Fill hot jars immediately with jelly, leaving a ¼-inch headspace. Wipe jar rims and adjust lids.

Process in a boiling water-bath canner.
Half-pints, 5 minutes

CRAB APPLE JELLY
Makes 5 or 6 half-pints.

3 pounds crab apples
3 cups water

4 cups sugar

Use firm, crisp crab apples, about one-fourth slightly underripe and three-fourths fully ripe. See General Steps in Making Jelly (pp. 126-129). Wash crab apples and discard any bad ones. Remove stems and blossom ends. Do not peel or core them. Cut crab apples into small pieces and discard seeds. Place apples in a large, thick-bottom saucepan. Add water, cover, and bring to a boil on high heat. Reduce heat and simmer 20 to 25 minutes or until apples are soft. Extract juice (pp. 126-127). Measure 4 cups crab apple juice into a saucepan. Add sugar and stir well. Bring to a boil on high heat. Boil rapidly until jelly is done (see Testing for Doneness, pp. 128-129). Remove from heat and quickly skim off foam. Fill hot jars immediately with jelly, leaving a ¼-inch headspace. Wipe jar rims and adjust lids.

Process in a boiling water-bath canner.
Half-pints, 5 minutes

GRAPE JELLY
Makes 3 or 4 half-pints.

3½ pounds grapes
½ cup water

3 cups sugar

Use about one-fourth slightly underripe and three-fourths fully ripe grapes. See General Steps in Making Jelly (pp. 126-129). Wash grapes and discard bad ones. Remove stems. Crush grapes in a large, thick-bottom saucepan. Add water, cover, and bring to a boil on high heat. Reduce heat and simmer 10 minutes. Extract juice (pp. 126-127). To prevent formation of tartrate crystals in the jelly, refrigerate juice overnight. Then, strain juice through two thicknesses of damp cheesecloth to remove crystals that have formed. Measure 4 cups grape juice into a saucepan. Add sugar and stir well. Bring to a boil on high heat. Boil rapidly until jelly is done (see Testing for Doneness, pp. 128-129). Remove jelly from heat and quickly skim off foam. Fill hot jars immediately with jelly, leaving a ¼-inch headspace. Wipe jar rims and adjust lids.

Process in a boiling water-bath canner.
Half-pints, 5 minutes

GRAPE PLUM JELLY
Makes about 10 half-pints.

3½ pounds plums
3 pounds Concord grapes
1 cup water

8½ cups sugar
1 1¾-ounce box powdered pectin

Use fully ripe plums and grapes. See General Steps in Making Jelly (pp. 126-129). Wash plums and grapes and remove seeds; do not peel them. Thoroughly crush plums and grapes, one layer at a time, in a large, thick-bottom saucepan. Add water and bring to a boil on high heat. Reduce heat, cover, and simmer 10 minutes. Extract juice (pp. 126-127). Measure sugar and set it aside. Then, measure 6½ cups juice into a saucepan. Add pectin. Bring to a rolling boil over high heat, stirring constantly. Add the sugar and return to a full rolling boil. Boil rapidly for 1 minute, stirring constantly. Remove from heat and quickly skim off foam. Fill hot jars immediately with jelly, leaving a ¼-inch headspace. Wipe jar rims and adjust lids.

Process in a boiling water-bath canner.
Half-pints, 5 minutes

LOQUAT JELLY
Makes 4 or 5 half-pints.

3 pounds loquats

4 cups sugar

Use full-size loquats that are still hard. See General Steps in Making Jelly (pp. 126-129). Wash loquats and remove the blossom ends. Place them in a large, thick-bottom saucepan and barely cover them with cold water. Cook slowly until the pulp is very soft. Extract juice (pp. 126-127) and place it in a saucepan. Cook it down until it is thick and cherry-colored. Then, measure 4 cups of juice into a saucepan. Add sugar and stir well. Bring to a boil on high heat. Boil rapidly until jelly is done (see Testing for Doneness, pp. 128-129). Remove jelly from heat and quickly skim off foam. Fill hot jars immediately with jelly, leaving a ¼-inch headspace. Wipe jar rims and adjust lids.

Process in a boiling water-bath canner.
Half-pints, 5 minutes

MAYHAW JELLY
Makes 2 half-pints.

1 pound mayhaws
2 cups water

1½ cups sugar

Use about one-fourth slightly underripe and three-fourths fully ripe mayhaws. See General Steps in Making Jelly (pp. 126-129). Wash mayhaws and place in a large, thick-bottom saucepan. Add water and bring to a boil on high heat. Reduce heat and simmer 10 to 15 minutes or until tender enough to mash. Extract juice (pp. 126-127). Measure 2 cups juice into a saucepan. Add sugar and stir until dissolved. Bring to a boil on high heat. Boil rapidly until jelly is done (see Testing for Doneness, pp. 128-129). Remove from heat and quickly skim off foam. Fill hot jars immediately with jelly, leaving a ¼-inch headspace. Wipe jar rims and adjust lids.

Process in a boiling water-bath canner.

Half-pints, 5 minutes

MINT JELLY
Makes 3 or 4 half-pints.

1½ cups firmly packed mint
2¼ cups water
Few drops green food coloring (optional)
3½ cups sugar
2 tablespoons lemon juice
6 ounces liquid pectin

Use fresh mint. See General Steps in Making Jelly (pp. 126-129). Wash mint and crush leaves and stems or chop them into very small pieces. Place mint in a large, thick-bottom saucepan and add water. Bring to a boil on high heat. Remove pan from heat, cover it, and let it sit for 10 minutes. Add food coloring. Strain and discard mint. Measure 1¾ cups mint juice into a saucepan. Add sugar and lemon juice and stir. Place on high heat, stir constantly, and bring to a full rolling boil that cannot be stirred down. Add liquid pectin and heat again to a full rolling boil. Boil rapidly for 1 minute. Remove from heat and quickly skim off foam. Fill hot jars immediately with jelly, leaving a ¼-inch headspace. Wipe jar rims and adjust lids.

Process in a boiling water-bath canner.

Half-pints, 5 minutes

MUSCADINE JELLY
Makes about 9 half-pints.

4 pounds muscadines
3 cups sugar

Use muscadines that are in the just-ripe stage. See General Steps in Making Jelly (pp. 126-129). Wash muscadines and remove stems; leave skins on. Crush muscadines in a large, thick-bottom saucepan. Do not add water. Bring to a boil on high heat. Reduce heat and simmer about 10 minutes, stirring constantly. Press juice from the hot grapes. Pour juice into large glass bottles or pitchers. Place in refrigerator. The next day, strain the juice through several thicknesses of cheesecloth. Measure 4 cups juice into a saucepan. Bring to a boil on high heat. Add sugar and stir until it dissolves. Boil rapidly until jelly is done (see Testing for Doneness, pp. 128-129). Remove from heat and quickly skim off foam. Fill hot jars immediately with jelly, leaving a ¼-inch headspace. Wipe jar rims and adjust lids.

Process in a boiling water-bath canner.

Half-pints, 5 minutes

PEACH JELLY
Makes 5 or 6 half-pints.

3½ pounds peaches
½ cup water
5 cups sugar
½ cup lemon juice
1 box powdered pectin

Use fully ripe peaches that are not damaged by insects or disease. See General Steps in Making Jelly (pp. 126-129). Wash and slice or chop peaches. Do not peel. Crush fruit and place it in a large, thick-bottom saucepan. Add water, cover, and bring to a boil on high heat. Reduce heat and simmer 5 minutes, stirring occasionally. Extract juice (pp. 126-127). Measure 3 cups juice into a saucepan. Measure sugar and set it aside. Add lemon juice and pectin to juice and bring to a full boil on high heat. Stir constantly. As soon as mixture boils, add sugar and stir to dissolve it. Bring mixture back to a full rolling boil that cannot be stirred down. Boil rapidly 1 minute, stirring constantly. Remove from heat and quickly skim off foam. Fill hot jars immediately, leaving a ¼-inch headspace. Wipe jar rims and adjust lids.

Process in a boiling water-bath canner.
Half-pints, 5 minutes

JALAPENO PEPPER JELLY
Makes 2 half-pints.

3 pounds jalapeno peppers
1 cup white vinegar (5% acidity)
5 cups sugar
6 ounces liquid pectin

See General Steps in Making Jelly (pp. 126-129). Wash peppers thoroughly. Remove stems and seeds. Run peppers through a food grinder or blender. Prepare enough to make 2 cups of pulp and juice. In a large, thick-bottom saucepan, combine pulp and juice with vinegar and sugar; stir until sugar is dissolved. Boil for 10 minutes over medium heat. Remove from heat. Add liquid pectin and increase heat to high; boil rapidly for 1 minute. Remove from heat and quickly skim off foam. Fill hot jars immediately with jelly, leaving a ¼-inch headspace. Wipe jar rims and adjust lids.

Process in a boiling water-bath canner.
Half-pints, 5 minutes

PLUM JELLY
Makes about 3 half-pints.

2 pounds plums
1 cup water
1½ cups sugar

See General Steps in Making Jelly (pp. 126-129). Wash plums and crush them. Place plums in a large, thick-bottom saucepan. Add water, cover, and bring to a boil on high heat. Then, reduce heat and simmer until plums are soft, 15 to 20 minutes. Extract juice (pp. 126-127). Measure 2 cups juice into a saucepan. Add sugar and stir well to dissolve it. Bring to a boil over high heat, stirring constantly. Boil rapidly until jelly is done (see Testing for Doneness, pp. 128-129). Remove from heat and quickly skim off foam. Fill hot jars immediately with jelly, leaving a ¼-inch headspace. Wipe jar rims and adjust lids.

Process in a boiling water-bath canner.
Half-pints, 5 minutes

STRAWBERRY RHUBARB JELLY
Makes about 7 half-pints.

1½ pounds red rhubarb stalks
1½ quarts ripe strawberries
6 cups sugar
6 ounces liquid pectin

Use young, tender, well-colored rhubarb stalks. Use firm-ripe strawberries. See General Steps in Making Jelly (pp. 126-129). Wash and cut rhubarb into 1-inch pieces; then, chop it in a blender or grinder. Wash strawberries and remove caps and stems. Crush berries, one layer at a time, in a bowl or saucepan. Place both rhubarb and berries in a jelly bag or double layer of cheesecloth and gently squeeze out juice. Measure 3½ cups combined juice into a large, thick-bottom saucepan. Add sugar and stir well to dissolve it. Bring to a boil over high heat, stirring constantly. Immediately stir in pectin. Then, bring to a full rolling boil and boil rapidly for 1 minute, stirring constantly. Remove from heat and quickly skim off foam. Fill hot jars immediately with jelly, leaving a ¼-inch headspace. Wipe jar rims and adjust lids.

Process in a boiling water-bath canner.

MAKING OTHER JELLIED SPREADS

Jams, Preserves, Marmalades, Conserves, Butters, and Syrup

Tips for Making Jellied Spreads

1. Read General Steps in Canning (pp. 19-20) for information on using a water-bath canner, preparing and testing jars and lids, filling jars, and processing.

2. If you are not following an exact recipe, you may want to test the pectin and acid in your fruit to be sure there is enough to gel. Use the same tests as in Testing Juice (pp. 127-128).

3. Ripe fruit is desirable for both flavor and texture. Don't use overripe fruits.

4. Fruits may be crushed, chopped, or ground in a food processor. A potato masher is useful for crushing.

5. Cook small batches at a time in a large, thick-bottom saucepan to help prevent scorching.

6. The proportion of sugar to fruit is generally ¾ to 1 cup sugar for each cup of prepared fruit.

7. Stir fruit and sugar over low heat until sugar dissolves. Then, boil rapidly on high heat for a clear finished product. Stir frequently to prevent sticking or scorching.

8. Adding ¼ teaspoon butter or margarine during cooking reduces the development of foam on the syrup.

9. To test for doneness, use the temperature test (p. 128) or the refrigerator-freezer test (p. 129) used for jellies.

10. Fill hot jars immediately and leave a ¼-inch headspace. Remove air bubbles, wipe jar rims, adjust lids, and process in a water-bath canner same as for jelly (see Canning Jelly, p. 129).

JAM RECIPES

Jams are generally made from small fruits, and the entire fruit is used. However, no effort is made to retain the shape of the fruit and a more or less homogeneous mixture results. Only the fleshy portions of large fruits are used.

BLACKBERRY JAM
Makes about 4 half-pints.

2 pounds blackberries   3 cups sugar

See Tips for Making Jellied Spreads (pp. 138-139). Wash blackberries and remove caps. Crush berries in a large, thick-bottom saucepan. Heat the berries. If you do not want seeds in the jam, press crushed berries through a fine sieve. Add sugar to berries in saucepan and heat slowly to a boil, stirring occasionally until sugar dissolves. Then, cook over high heat until it is done (see Testing for Doneness, pp. 128-129). Fill hot jars immediately with jam, leaving a ¼-inch headspace. Remove air bubbles. Wipe jar rims and adjust lids.

Process in a boiling water-bath canner.
Half-pints, 5 minutes
Pints and Quarts, 10 minutes

BLUEBERRY SPICE JAM
Makes about 5 half-pints.

2½ pints ripe blueberries
1 tablespoon lemon juice
½ teaspoon ground nutmeg or cinnamon

5½ cups sugar
¾ cup water
1 1¾-ounce box powdered pectin

See Tips for Making Jellied Spreads (pp. 138-139). Wash and thoroughly crush blueberries, one layer at a time, in a large, thick-bottom saucepan. Add lemon juice, nutmeg, sugar, and water. Bring to a boil on high heat and immediately stir in pectin. Heat to a full rolling boil and boil rapidly for 1 minute, stirring constantly. Remove from heat and quickly skim off foam. Fill hot jars immediately with jam, leaving a ¼-inch headspace. Remove air bubbles. Wipe jar rims and adjust lids.

Process in a boiling water-bath canner.
Half-pints, 5 minutes

FIG JAM
Makes about 6 half-pints.

3 pounds figs
6 quarts boiling water

3 cups sugar
3 tablespoons lemon juice

Use firm, ripe figs. See Tips for Making Jellied Spreads (pp. 138-139). Wash figs and remove stems. Place figs in a large baking pan and cover them with boiling water. Let them sit for 5 minutes. Drain and rinse in clean, cold water. Grind figs through a food processor, using a medium blade, and place in a thick-bottom saucepan. Add sugar and lemon juice. Heat slowly to a boil, stirring to dissolve sugar. Then, increase heat and boil rapidly until mixture is quite thick. Stir frequently to prevent sticking. Fill hot jars immediately with jam, leaving a ¼-inch headspace. Remove air bubbles. Wipe jar rims and adjust lids.

Process in a boiling water-bath canner.
Half-pints, 5 minutes
Pints and Quarts, 10 minutes
Note: Do not try to cook more than 3 pounds of figs at a time.

GRAPE JAM
Makes about 3 half-pints.

1 pound grapes (after removing stems and seeds)   1 cup sugar

See Tips for Making Jellied Spreads (pp. 138-139). Wash grapes, remove stems, and press pulp from skins. Save skins. Place pulp in a saucepan and boil until tender. Then, press pulp through a sieve to remove seeds. Chop skins or grind them through a food processor, using a medium blade. Add seedless pulp and weigh it. Place 1 pound of the mixture in a large, thick-bottom saucepan. Heat slowly to a boil, add sugar, and cook until skins are tender, stirring frequently. Then, increase heat and boil rapidly until thick. Fill hot jars immediately with jam, leaving a ¼-inch headspace. Remove air bubbles. Wipe jar rims and adjust lids.

Process in a boiling water-bath canner.
Half-pints, 5 minutes
Pints and Quarts, 10 minutes

Note: If using a sweet variety of grapes, you may need to use less sugar. For highly acid grapes, increase the sugar up to ¾ pound for each pound of grapes.

PEACH JAM
Makes about 4 half-pints.

3 pounds peaches · · · · · · · · · · · · · 3 cups sugar

See Tips for Making Jellied Spreads (pp. 138-139). Wash and peel peaches; remove seeds. Grind peaches in a food processor, using a medium blade. Weigh 2 pounds ground peaches and place in a large, thick-bottom saucepan. Add sugar. Heat slowly to a boil, stirring continually until sugar dissolves. Then, increase heat and boil rapidly until mixture is somewhat thick. Fill hot jars immediately with jam, leaving a ¼-inch headspace. Remove air bubbles. Wipe jar rims and adjust lids.

Process in a boiling water-bath canner.
Half-pints, 5 minutes
Pints, 10 minutes

PEAR JAM
Makes about 3 half-pints.

1 pound pears · · · · · · · · · · · · · · · ¼ cup water
1½ cups sugar · · · · · · · · · · · · · · · 1 teaspoon lemon juice

See Tips for Making Jellied Spreads (pp. 138-139). Wash, peel, and core pears. Put them through a food processor, using a coarse blade. Prepare enough to make 1½ cups ground pears. Place pears in a large, thick-bottom saucepan. Add sugar, water, and lemon juice. Heat slowly to a boil, stirring until sugar is dissolved. Then, increase heat and boil rapidly until mixture is thick and clear. Fill hot jars immediately with jam, leaving a ¼-inch headspace. Remove air bubbles. Wipe jar rims and adjust lids.

Process in a boiling water-bath canner.
Half-pints, 5 minutes
Pints and Quarts, 10 minutes

PEAR APPLE JAM
Makes about 7 half-pints.

2 pounds pears	6½ cups sugar
1 pound apples	⅓ cup bottled lemon juice
¼ teaspoon ground cinnamon	6 ounces liquid pectin

See Tips for Making Jellied Spreads (pp. 138-139). Wash, peel, and core pears and apples. Chop them into very small pieces. Prepare enough to make 2 cups chopped pears and 1 cup chopped apples. Crush pears and apples in a large, thick-bottom saucepan, add cinnamon, and mix well. Add sugar and lemon juice and stir to thoroughly mix. Bring to a boil over high heat, stirring constantly. Immediately stir in pectin. Then, bring to a rolling boil and boil for 1 minute, stirring constantly. Remove from heat and quickly skim off foam. Fill hot jars immediately with jam, leaving a ¼-inch headspace. Remove air bubbles. Wipe jar rims and adjust lids.

Process in a boiling water-bath canner.
Half-pints, 5 minutes

PLUM JAM
Makes about 4 half-pints.

2 pounds plums	3 cups sugar
¾ cup water	2 tablespoons lemon juice

See Tips for Making Jellied Spreads (pp. 138-139). Wash plums and chop them. Place plums in a large, thick-bottom saucepan. Add water, sugar, and lemon juice. Heat slowly to a boil and cook until sugar dissolves, stirring frequently. Then, increase heat and boil rapidly about 15 minutes. Fill hot jars immediately with jam, leaving a ¼-inch headspace. Remove air bubbles. Wipe jar rims and adjust lids.

Process in a boiling water-bath canner.
Half-pints, 5 minutes
Pints and Quarts, 10 minutes

STRAWBERRY JAM
Makes about 4 half-pints.

2 pounds strawberries	3 cups sugar

See Tips for Making Jellied Spreads (pp. 138-139). Wash berries and remove caps. Crush berries in a large, thick-bottom saucepan and add sugar. Heat slowly to a boil, stirring occasionally until sugar dissolves. Then, increase heat and boil rapidly until mixture is thick, stirring frequently to prevent sticking. Fill hot jars immediately with jam, leaving a ¼-inch headspace. Remove air bubbles. Wipe jar rims and adjust lids.

Process in a boiling water-bath canner.
Half-pints, 5 minutes
Pints and Quarts, 10 minutes

PRESERVES RECIPES

A fruit preserve consists of whole, small fruits, or pieces of large fruits, cooked in a syrup until clear and stored in the syrup or the jellied juice. The ideal preserve retains the color and flavor of the fresh fruit.

FIG PRESERVES
Makes 4 to 5 half-pints.

1½ quarts figs	3 cups water
6 cups boiling water	1 lemon, thinly sliced
2 cups sugar	

Use firm, ripe figs. See Tips for Making Jellied Spreads (pp. 138-139). Wash figs. Place them in a large baking pan and cover with boiling water. Let them sit 15 minutes; then, drain and rinse in cold water. Mix sugar and 3 cups water in a saucepan. Add lemon slices. Heat to boiling over high heat and boil 10 minutes to make a syrup. Remove from heat and quickly skim off foam. Remove lemon slices. Return to high heat and drop figs into syrup, a few at a time so syrup will continue to boil. Boil rapidly until figs are transparent. Remove figs and place them in a shallow pan. Continue boiling the syrup until it is thick. Then, pour syrup over figs and let them sit for 6 to 8 hours. Reheat figs and syrup to boiling. Fill hot jars with figs, leaving a ¼-inch headspace. Add syrup to cover figs, leaving a ¼-inch headspace. Remove air bubbles. Wipe jar rims and adjust lids.

Process in a boiling water-bath canner.
Half-pints, 5 minutes

MUSCADINE PRESERVES
Makes about 10 half-pints.

5 pounds muscadines	10 cups sugar

See Tips for Making Jellied Spreads (pp. 138-139). Wash muscadines. Rub them through ½-inch hardware cloth to separate pulp from hulls. Save hulls. Place pulp and juice in a large, thick-bottom saucepan and simmer until pulp is broken down enough to free the seeds (about 5 minutes). Force pulp through a colander and discard seeds. Cook hulls and pulp over low heat until hulls are tender, about 15 minutes. Time will vary with muscadine variety. Add sugar to simmering mixture and stir to dissolve sugar. Cook until mixture is done (see Testing for Doneness, pp. 128-129). Fill hot jars immediately with preserves, leaving a ¼-inch headspace. Remove air bubbles. Wipe jar rims and adjust lids.

Process in a boiling water-bath canner.
Half-pints, 5 minutes
Pints and Quarts, 10 minutes

To Vary: To make muscadine jam, grind or chop hulls before adding the pulp and sugar.

PEACH PRESERVES
Makes about 7 half-pints.

| 10 large peaches | 6 cups sugar |

Use hard, ripe peaches. See Tips for Making Jellied Spreads (pp. 138-139). Wash peaches. Dip them in boiling water for 30 to 60 seconds until skins loosen. Then, dip them quickly in cold water and slip off skins. Cut in half, remove pits, and cut them in slices. Combine peaches and sugar in a large, thick-bottom saucepan and let them sit 12 to 18 hours in a cool place. Then, heat peach mixture slowly to a boil, stirring often to prevent sticking. Boil gently until fruit is clear and syrup is thick, about 40 minutes. As mixture thickens, stir frequently to prevent sticking. Remove from heat and quickly skim off foam, if necessary. Fill hot jars immediately with peaches, leaving a ¼-inch headspace. Add boiling syrup to cover peaches, leaving a ¼-inch headspace. Remove air bubbles. Wipe jar rims and adjust lids.

Process in a boiling water-bath canner.
Half-pints, 5 minutes

PEAR PRESERVES
Makes about 5 half-pints.

6 medium-size pears   2½ cups water
1 lemon   1½ cups sugar
1½ cups sugar

Use hard, ripe pears. See Tips for Making Jellied Spreads (pp. 138-139). Wash pears and peel them. Cut them in halves or quarters and remove cores. Wash lemon and cut it into thin slices. Combine 1½ cups sugar and water in a large, thick-bottom saucepan. Heat over high heat and boil 2 minutes. Add pears, reduce heat, and boil gently for 15 minutes. Add 1½ cups sugar and lemon slices. Stir until sugar dissolves. Then, increase heat and boil rapidly until pears are clear, about 25 minutes. Remove pan from heat, cover it, and let pears sit for 12 to 24 hours in a cool place. Then, heat pears and syrup to boiling. Fill hot jars immediately with pears, leaving a ½-inch headspace. Add boiling syrup to cover pears, leaving a ¼-inch headspace. Remove air bubbles. Wipe jar rims and adjust lids.

Process in a boiling water-bath canner.
Half-pints, 5 minutes.

GINGERED PEAR PRESERVES
Makes about 4 half-pints.

2 pounds pears
1½ cups sugar
½ to 1 cup water
2 pieces gingerroot
Rind of ½ lemon

See Tips for Making Jellied Spreads (pp. 138-139). Use hard, ripe pears. Peel and core them and cut into uniform-size pieces. Place sugar and water in a large, thick-bottom saucepan. Bring to a boil over high heat and boil 2 minutes, stirring some to dissolve sugar. Add pears, reduce heat, and boil gently for 15 minutes, stirring occasionally. Increase heat and boil rapidly until pears are clear, about 25 minutes. In a small saucepan, cook gingerroot and lemon rind in a small amount of water until they are tender. Drain and add them to pears; boil rapidly until syrup is somewhat thick. Fill hot jars immediately with pears, leaving a ½-inch headspace. Add syrup to cover pears, leaving a ¼-inch headspace. Remove air bubbles. Wipe jar rims and adjust lids.

Process in a boiling water-bath canner.
Half-pints, 5 minutes
Pints and Quarts, 10 minutes

Note: You can let pears plump overnight in the syrup before filling jars. Then, reheat and fill hot jars, leaving a ½-inch headspace. Process for 25 minutes in pints and quarts.

PLUM PRESERVES
Makes about 4 half-pints.

2 pounds plums
3 to 4 cups sugar
1 cup water

See Tips for Making Jellied Spreads (pp. 138-139). Wash plums, cut in half, and remove pits. Combine sugar and water in a large, thick-bottom saucepan. Heat slowly to a boil, stirring to dissolve sugar. Add plums and boil gently about 15 minutes or until fruit is tender. Stir frequently to prevent sticking. Fill hot jars immediately with plums, leaving a ¼-inch headspace. Add syrup to cover plums, leaving a ¼-inch headspace. Remove air bubbles. Wipe jar rims and adjust lids.

Process in a boiling water-bath canner.
Half-pints, 5 minutes
Pints and Quarts, 10 minutes

Note: You can let plums plump in syrup for several hours before filling jars. Then, reheat and fill hot jars, leaving a ¼-inch headspace. Process for 25 minutes in pints and quarts.

PUMPKIN PRESERVES
Makes about 4 half-pints.

1 pound pumpkin strips (see p. 82 for preparing pumpkin)

2 cups sugar
¼ cup lemon juice

See Tips for Making Jellied Spreads (pp. 138-139). Cut pumpkin strips into thin shavings. Spread the shavings on a platter, sprinkle with sugar and lemon juice, and let them sit 8 to 10 hours. Place pumpkin mixture into a large, thick-bottom saucepan and cook over low heat until it is clear and tender, about 1½ hours. Stir occasionally to prevent sticking. Remove from heat and quickly skim off foam. Fill hot jars immediately with pumpkin mixture, leaving a ¼-inch headspace. Remove air bubbles. Wipe jar rims and adjust lids.

Process in a boiling water-bath canner.
Half-pints, 5 minutes
Pints, 10 minutes

QUINCE PRESERVES
Makes about 4 half-pints.

3 pounds quince
2 quarts water

3 cups sugar

See Tips for Making Jellied Spreads (pp. 138-139). Wash quince and remove stems and blossom ends. Peel, cut them in quarters, remove cores, and discard all gritty parts. Combine sugar and water in a large, thick-bottom saucepan and boil 5 minutes. Add quince and cook until it has a clear, red color and the syrup almost reaches jellying point, about 1 hour. As syrup thickens, stir frequently to prevent sticking. Fill hot jars immediately with quince, leaving a ¼-inch headspace. Add syrup to cover quince, leaving a ¼-inch headspace. Remove air bubbles. Wipe jar rims and adjust lids.

Process in a boiling water-bath canner.
Half-pints, 5 minutes
Pints and Quarts, 10 minutes

STRAWBERRY PRESERVES
Makes about 2 half-pints.

1 pound strawberries

2 cups sugar

Use berries that are red, firm, and ripe; do not use berries with hollow cores. See Tips for Making Jellied Spreads (pp. 138-139). Sort berries according to size; wash them and remove stems and caps. Combine berries and sugar in a large bowl and let them sit 8 to 10 hours. Place them in a large, thick-bottom saucepan and heat slowly to a gentle boil, stirring to dissolve sugar. Then, increase heat and boil rapidly until berries are clear. Remove berries and cook syrup until it reaches the desired thickness. Remove pan from heat and quickly skim off foam. Add berries to syrup. Fill hot jars immediately with preserves, leaving a ¼-inch headspace. Remove air bubbles. Wipe jar rims and adjust lids.

Process in a boiling water-bath canner.
Half-pints, 5 minutes
Pints and Quarts, 10 minutes

Note: You can pour cooked mixture into a flat pan and allow berries to plump for 8 to 10 hours. Then, reheat berries to simmering and fill hot jars, leaving a ¼-inch headspace. Process for 25 minutes in pints or quarts.

STRAWBERRY FIG PRESERVES
Makes about 6 half-pints.

2 quarts figs
2 3-ounce packages strawberry gelatin
3 cups sugar

See Tips for Making Jellied Spreads (pp. 138-139). Wash, peel, and mash figs. Measure 3 cups mashed figs into a large, thick-bottom saucepan. Add gelatin and sugar and heat to a boil. Lower heat and boil slowly for 3 to 5 minutes, stirring often. Fill hot jars immediately with fig mixture, leaving a ¼-inch headspace. Remove air bubbles. Wipe jar rims and adjust lids.

Process in a boiling water-bath canner.
Half-pints, 5 minutes

WATERMELON RIND PRESERVES
Makes about 3 half-pints.

1 pound watermelon rind
2 tablespoons salt
Water
2 cups sugar
2 tablespoons gingerroot
½ lemon, thinly sliced

Use only the white part of the rind. See Tips for Making Jellied Spreads (pp. 138-139). Cut rind into 1-inch pieces. Dissolve salt in 1 quart water. Soak pieces in salt water for 5 to 6 hours. Drain, rinse, and drain again. Soak rind in 1 quart clear water for 30 minutes. Then, drain. Place rind in a large, thick-bottom saucepan and add 1 quart fresh water. Boil for about 1½ hours. Drain again. Combine 2 quarts fresh water and sugar in a saucepan. Heat to boiling, stirring to dissolve sugar. Boil 5 minutes. Drop watermelon rind and gingerroot into boiling syrup. Boil for about 1 hour. As syrup thickens, add lemon slices. Continue to boil until syrup is somewhat thick. Fill hot jars immediately with rind, leaving a ¼-inch headspace. Add syrup to cover rind, leaving a ¼-inch headspace. Remove air bubbles. Wipe jar rims and adjust lids.

Process in a boiling water-bath canner.
Half-pints, 5 minutes
Pints and Quarts, 10 minutes

MARMALADE RECIPES

Marmalades have the combined characteristics of jellies and preserves. They are prepared from pulpy fruit, preferably those that contain pectin. The pulp and, if desired, the skin of the fruit are suspended in the jellied juice. Citrus fruits are especially desirable for pectin content as well as for flavor. Pectin is extracted more rapidly by cooking fruit before sugar is added. Fruit is added in distinct slices or shreds and is cooked until clear.

AMBER MARMALADE
Makes about 2 half-pints.

1 orange	Water
1 grapefruit	Sugar
1 lemon	⅛ teaspoon canning salt

Use smooth, thick-skinned, well-colored fruit that is free of blemishes. See Tips for Making Jellied Spreads (pp. 138-139). Peel orange, grapefruit, and lemon and remove white membrane under peeling. Set fruit pulp aside. Slice peeling into very thin strips. Parboil strips of peeling three times by adding 1 quart cold water to strips in a large, thick-bottom saucepan, bringing it to a boil, and boiling 5 minutes. Then, drain and discard water after each cooking. Cut pulp of fruit into thin slices and remove membranes. Combine pulp and drained, parboiled peel. Weigh or measure fruit (pulp, juice, and parboiled peel) into a large saucepan. Add 2 cups water for each cup of fruit mixture. Heat to a boil on high heat and boil rapidly for 40 minutes. Weigh or measure fruit again and add 1 cup sugar for each cup of fruit. Add salt. Boil mixture on high heat for 25 minutes or until it is done (see Testing for Doneness, pp. 128-129). If a thick marmalade is desired, boil longer but stir and watch carefully to avoid scorching. Remove from heat and let stand until slightly cool. Then, stir to distribute fruit equally. Fill hot jars immediately with marmalade, leaving a ¼-inch headspace. Remove air bubbles. Wipe jar rims and adjust lids.

Process in a boiling water-bath canner.
Half-pints, 5 minutes
Pints, 15 minutes

CARROT ORANGE MARMALADE
Makes about 3 half-pints.

6 medium-size carrots	Sugar
3 oranges	Juice and grated rind of 1 lemon

See Tips for Making Jellied Spreads (pp. 138-139). Wash carrots and oranges well. Remove tops from carrots. Wash and grind carrots in a food processor, using a medium blade. Cut oranges in half, discard seeds, and grind peel and pulp. Add oranges to carrots. Measure mixture into a large, thick-bottom saucepan. Add ⅔ cup sugar for each cup of mixture. Add lemon juice and grated lemon rind. Bring to a boil on high heat and boil rapidly until mixture is clear. Fill hot jars immediately with marmalade, leaving a ¼-inch headspace. Remove air bubbles. Wipe jar rims and adjust lids.

Process in a boiling water-bath canner.
Half-pints, 5 minutes
Pints, 15 minutes

CITRUS FRUIT MARMALADE
Makes about 4 pints.

1 grapefruit	6 quarts cold water
2 lemons	7 pounds (14 cups) sugar
6 oranges	

See Tips for Making Jellied Spreads (pp. 138-139). Wash fruits and cut them in half. Remove seeds. Grind pulp and peel in a food processor, using a medium blade. Cover ground fruit with cold water and let it sit overnight. Next morning, place fruit mixture in a large, thick-bottom saucepan. Heat it slowly to a boil. Then, increase heat and boil rapidly until peel is tender. Add sugar and boil mixture over high heat until it is done (see Testing for Doneness, pp. 128-129). Let mixture cool some. Fill hot jars with slightly cooled marmalade, leaving a ¼-inch headspace. Remove air bubbles. Wipe jar rims and adjust lids.

Process in a boiling water-bath canner.
Half-pints, 5 minutes
Pints, 15 minutes

GRAPE MARMALADE

Use about one-half slightly underripe and one-half fully ripe grapes. See Tips for Making Jellied Spreads (pp. 138-139). Wash grapes and remove stems. Press grapes to separate pulp from skins. Place pulp in a large, thick-bottom saucepan and cook 10 minutes. Then, press pulp through a sieve to remove seeds. Measure skins into a saucepan and add ¾ cup water for each quart of skins. Boil until tender. Mix pulp and skins together and measure mixture into a saucepan. Heat mixture to a boil. Add 1 pound sugar for each quart of grape mixture and stir to dissolve sugar. Then, increase heat and boil rapidly, stirring frequently until it is done (see Testing for Doneness, pp. 128-129). Fill hot jars immediately with marmalade, leaving a ¼-inch headspace. Remove air bubbles. Wipe jar rims and adjust lids.

Process in a boiling water-bath canner.
Half-pints, 5 minutes
Pints and Quarts, 10 minutes

GREEN TOMATO MARMALADE
Makes about 4 pints.

4 pounds green tomatoes	1 cup water
1 lemon	4 cups sugar
2 oranges	½ teaspoon salt

See Tips for Making Jellied Spreads (pp. 138-139). Wash tomatoes, lemon, and oranges. Remove stem ends of tomatoes. Cut tomatoes into small pieces. Cut lemon and oranges into thin slices; place them in a large, thick-bottom saucepan and add water. Cook over medium heat until tender. Add tomatoes, sugar, and salt. Cook over medium heat, stirring to dissolve sugar. Then, increase heat and boil rapidly until tomatoes are soft and mixture has thickened somewhat. Fill hot jars immediately with marmalade, leaving a ¼-inch headspace. Remove air bubbles. Wipe jar rims and adjust lids.

Process in a boiling water-bath canner.
Half-pints, 5 minutes
Pints, 10 minutes

KUMQUAT MARMALADE
Makes 4 to 5 half-pints.

1 quart kumquats	2¼ cups sugar
Water	

Use good, mature kumquats that have no insect or disease damage. See Tips for Making Jellied Spreads (pp. 138-139). Wash kumquats thoroughly. Cut each fruit in half; remove and discard seeds. Put them through a food processor, using a medium blade. Measure kumquats into a saucepan. Add 3 cups water for each cup of fruit. Heat to a boil and boil 15 minutes. Then, cover and let them sit overnight. Next day, measure 3 cups of kumquat mixture into a large saucepan. Heat to a boil. Add sugar and boil rapidly, stirring until sugar is dissolved. Cook until mixture reaches 220°F on a candy thermometer. Remove from heat and let it sit until slightly cool. Then, stir to distribute fruit equally. Fill hot jars with kumquat mixture, leaving a ½-inch headspace. Remove air bubbles. Wipe jar rims and adjust lids.

Process in a boiling water-bath canner.
Half-pints, 10 minutes

ORANGE LEMON MARMALADE
Makes about 4 half-pints.

4 oranges	Sugar
4 lemons	¼ teaspoon canning salt
Water	

Use smooth, thick-skinned, well-colored fruit that is free of blemishes. See Tips for Making Jellied Spreads (pp. 138-139). Peel oranges and lemons and remove white membrane under peeling. Set fruit pulp aside. Slice peeling into very thin strips. Parboil strips of peeling three times by adding 1 quart cold water to strips in a large, thick-bottom saucepan, bringing it to a boil, and boiling 5 minutes. Then, drain and discard water after each cooking. Cut pulp of fruit into thin slices and remove membranes. Combine pulp and drained, parboiled peel. Measure fruit (pulp, juice, and parboiled peel) into a large saucepan. Add 3 cups water for each cup of fruit mixture. Boil mixture rapidly 40 minutes. Measure fruit mixture again and add 1 cup sugar for each cup of fruit. Add salt. Boil mixture on high heat for 15 minutes or until it is done (see Testing for Doneness, pp. 128-129). If a thicker marmalade is desired, boil longer but stir and watch carefully to avoid scorching. Remove from heat and let stand until slightly cool. Then, stir to distribute fruit equally. Fill hot jars immediately with marmalade, leaving a ¼-inch headspace. Remove air bubbles. Wipe jar rims and adjust lids.

Process in a boiling water-bath canner.
Half-pints, 5 minutes
Pints, 15 minutes

PEAR MARMALADE
Makes about 2 half-pints.

2 pounds pears	4½ cups sugar
2 oranges	1 cup water

See Tips for Making Jellied Spreads (pp. 138-139). Wash and cut pears into small strips or pieces. Peel oranges and remove the white membrane under the peel. Cut orange pulp into thin slices and discard seeds and membranes. Chop half of the peel into small pieces. Combine pears, oranges, sugar, and water in a large, thick-bottom saucepan. Bring to a boil on high heat and boil rapidly until mixture is thick and transparent. Stir frequently as it thickens. Fill hot jars immediately with marmalade, leaving a ¼-inch headspace. Remove air bubbles. Wipe jar rims and adjust lids.

Process in a boiling water-bath canner.
Half-pints, 5 minutes

PEAR HONEY MARMALADE
Makes about 2 half-pints.

2 pounds pears
½ cup canned crushed pineapple
1 teaspoon lemon juice
3 cups sugar

See Tips for Making Jellied Spreads (pp. 138-139). Wash, peel, and core pears. Put them through a food processor, using a coarse blade. Prepare enough to make 3 cups ground pears. Place pears in a large, thick-bottom saucepan. Add pineapple, lemon juice, and sugar. Heat slowly to a boil, stirring to dissolve sugar. Then, increase heat and boil rapidly until mixture is thick and clear. Fill hot jars immediately with marmalade, leaving a ¼-inch headspace. Remove air bubbles. Wipe jar rims and adjust lids.

Process in a boiling water-bath canner.
Half-pints, 5 minutes
Pints and Quarts, 10 minutes

PERSIMMON MARMALADE
Makes about 4 half-pints.

1 quart persimmons
1 cup water
Juice of 1 lemon
Sugar

See Tips for Making Jellied Spreads (pp. 138-139). Wash and peel persimmons; remove and discard the seeds. Combine persimmons and water in a large, thick-bottom saucepan. Boil until it makes a thick pulp. Add lemon juice. Measure pulp and add 1 cup sugar for each quart pulp. Boil 10 minutes longer. Fill hot jars immediately with marmalade, leaving a ¼-inch headspace. Remove air bubbles. Wipe jar rims and adjust lids.

Process in a boiling water-bath canner.
Half-pints, 5 minutes
Pints and Quarts, 10 minutes

CONSERVE RECIPES

Conserves are similar to jams but are made from a mixture of fruits. They usually contain nuts and sometimes raisins.

CRANBERRY CONSERVE
Makes about 4 half-pints.

1 orange
1 quart cranberries
2 cups water

3 cups sugar
½ cup seedless raisins
½ cup chopped pecans

See Tips for Making Jellied Spreads (pp. 138-139). Wash orange well and do not peel it, but chop it into very small pieces. Remove and discard seeds. Wash cranberries. Combine chopped orange and water in a large, thick-bottom saucepan. Cook over high heat until peel is tender, about 20 minutes. Reduce heat and add cranberries, sugar, and raisins. Return to a gentle boil, stirring occasionally until sugar dissolves. Then, increase heat and boil rapidly about 8 minutes or until mixture is almost done (see Testing for Doneness, pp. 128-129). As mixture thickens, stir frequently to prevent sticking. Add pecans the last 5 minutes of cooking. Fill hot jars immediately with conserve, leaving a ¼-inch headspace. Remove air bubbles. Wipe jar rims and adjust lids.

Process in a boiling water-bath canner.
Half-pints, 5 minutes
Pints, 10 minutes

DAMSON PLUM CONSERVE
Makes about 6 half-pints.

4 pounds plums
2 oranges
6 cups sugar

1 pound raisins
1 pound chopped pecans

See Tips for Making Jellied Spreads (pp. 138-139). Wash plums, cut them in half, remove seeds, and chop plums. Wash oranges and peel them. Remove white membrane under peel. Slice half of the peel into thin strips. Discard remaining peel and seeds. Slice orange pulp and remove membranes. Mix chopped plums, sliced peel, orange pulp, sugar, and raisins in a large, thick-bottom saucepan. Heat slowly to a boil, stirring to dissolve sugar. Then, increase heat and boil rapidly, stirring frequently, until mixture is bright and thick as jam. Add pecans 5 minutes before removing pan from heat. Fill hot jars immediately with conserve, leaving a ¼-inch headspace. Remove air bubbles. Wipe jar rims and adjust lids.

Process in a boiling water-bath canner.
Half-pints, 5 minutes
Pints, 10 minutes

FIG CONSERVE
Makes about 4 half-pints.

2 pounds figs
3 quarts boiling water
½ pound raisins
Juice of 1 lemon
1 orange
3 cups sugar
½ cup chopped pecans

Use fresh, firm, ripe figs. See Tips for Making Jellied Spreads (pp. 138-139). Wash figs and place them in a large baking pan. Cover with boiling water and let stand 5 minutes. Drain and rinse in clean, cold water. Cut figs and raisins in small pieces and place in a large, thick-bottom saucepan. Add lemon juice. Wash and peel orange, remove white membrane under peel, and cut peel into thin strips. Slice orange pulp and remove membranes. Add orange peel and pulp and sugar to fig mixture. Heat slowly to a boil, stirring to dissolve sugar. Then, increase heat and boil rapidly until mixture is thick and somewhat transparent, about 1 hour. Add pecans the last 5 minutes of cooking. Fill hot jars immediately with conserve, leaving a ¼-inch headspace. Remove air bubbles. Wipe jar rims and adjust lids.

Process in a boiling water-bath canner.
Half-pints, 5 minutes
Pints, 10 minutes

Note: You can use 1 quart canned figs in place of fresh figs. Canned figs will not need to be soaked in boiling water.

GRAPE CONSERVE
Makes about 3 half-pints.

1 pound grapes
¼ orange
1 cup sugar
¼ cup raisins
¼ teaspoon canning salt
¼ cup chopped walnuts

See Tips for Making Jellied Spreads (pp. 138-139). Wash grapes and remove stems. Slip off skins and place pulp in a large, thick-bottom saucepan. Chop skins and set them aside. Wash orange and chop it into very small pieces; discard seeds. Cook grape pulp about 10 minutes or until seeds show. Press pulp through a sieve to remove seeds. Return pulp to saucepan and add orange, sugar, raisins, and salt. Heat slowly to a boil, stirring to dissolve sugar. Then, increase heat and boil rapidly until mixture begins to thicken. Stir frequently to prevent sticking. Add chopped grape skins and cook 10 minutes or until mixture is somewhat thick. Stir in chopped walnuts the last 5 minutes of cooking. Fill hot jars immediately with conserve, leaving a ¼-inch headspace. Remove air bubbles. Wipe jar rims and adjust lids.

Process in a boiling water-bath canner.
Half-pints, 5 minutes
Pints, 10 minutes

PEAR CONSERVE
Makes about 4 half-pints.

4 cups coarsely ground pears
Juice and grated rind of 1 lemon
1 cup chopped raisins
3½ cups sugar
½ cup chopped pecans

See Tips for Making Jellied Spreads (pp. 138-139). Combine pears, lemon juice and rind, raisins, and sugar in a large, thick-bottom saucepan. Heat slowly to a boil, stirring to dissolve sugar. Then, increase heat and boil rapidly until mixture is thick and fruit is clear. Stir frequently to prevent sticking. Add pecans the last 5 minutes of cooking. Fill hot jars immediately with conserve, leaving a ¼-inch headspace. Remove air bubbles. Wipe jar rims and adjust lids.

Process in a boiling water-bath canner.
Half-pints, 5 minutes
Pints, 10 minutes

PEAR MUSCADINE CONSERVE
Makes about 6 half-pints.

3 pounds pears
1 pint muscadine juice
Sugar
1 cup chopped pecans

See Tips for Making Jellied Spreads (pp. 138-139). Wash, peel, and core pears. Grind them in a food processor, using a coarse blade. Prepare enough to make 4 cups of ground pears. Combine pears and muscadine juice in a large, thick-bottom saucepan. Simmer until pears are tender. Measure mixture; add 1 cup sugar for each cup of mixture. Cook over high heat, stirring frequently, until it is done (see Testing for Doneness, pp. 128-129). Add pecans the last 5 minutes of cooking. Fill hot jars immediately with hot conserve, leaving a ¼-inch headspace. Remove air bubbles. Wipe jar rims and adjust lids.

Process in a boiling water-bath canner.
Half-pints, 5 minutes
Pints, 10 minutes

FRUIT BUTTER RECIPES

Fruit butters are made by cooking the pulp of any fruit to a smooth consistency, thick enough to hold its shape and soft enough to spread easily. Fruits most commonly used for butters are tart apples, crab apples, grapes, peaches, pears, plums, and quince. You can use the residue left after extracting juice for jelly as a source of pulp for butters. If you do, add spice or acid for better flavor and gelling quality.

Testing Fruit Butter For Doneness. Remove a spoonful of hot butter from saucepan; hold it away from steam 2 minutes. If it remains in a mound on the spoon, it is done. Or, spoon a small amount onto a plate. If liquid separates from butter and forms a rim around the edge, cook it longer.

APPLE BUTTER
Makes about 9 half-pints.

8 pounds apples
2 cups apple cider
2 cups vinegar (5% acidity)
2¼ cups sugar
2¼ cups packed brown sugar
2 tablespoons ground cinnamon
1 tablespoon ground cloves

See Tips for Making Jellied Spreads (pp. 138-139). Wash apples and remove stems; cut apples in quarters and remove the cores. Place apples in a large, thick-bottom saucepan and add cider and vinegar. Cook slowly over medium heat until soft. Press pieces through a colander, food mill, or strainer. Return apple pulp to pan and add sugar, brown sugar, cinnamon, and cloves. Cook on medium heat, stirring constantly to prevent scorching. Cook until butter is thick and has taken on a glossiness or sheen. See Testing Fruit Butter for Doneness (above). Fill hot jars immediately with the butter, leaving a ¼-inch headspace. Remove air bubbles. Wipe jar rims and adjust lids.

Process in a boiling water-bath canner.
Half-pints, 5 minutes
Pints, 10 minutes

CRAB APPLE BUTTER
Makes about 4 half-pints.

2 pounds crab apples
1 cup water
2 cups sugar
3 cinnamon sticks, crushed
6 whole cloves

See Tips for Making Jellied Spreads (pp. 138-139). Wash crab apples, remove stems and blossom ends, and cut them in quarters. Place apples and water in a large, thick-bottom saucepan and cook until fruit is soft, stirring constantly. Put crab apples through a sieve. Measure pulp. Put 4 cups in a saucepan and add sugar. Tie cinnamon and cloves in a spice bag and add to mixture. Cook over high heat, stirring constantly to prevent scorching. As cooking progresses, reduce heat a little to prevent spattering. Cook until butter is thick and has taken on a glossiness or sheen. See Testing Fruit Butter for Doneness (above). Fill hot jars immediately with the butter, leaving a ¼-inch headspace. Remove air bubbles. Wipe jar rims and adjust lids.

Process in a boiling water-bath canner.
Half-pints, 5 minutes
Pints, 10 minutes

GRAPE BUTTER
Makes about 4 half-pints.

3 pounds grapes
2½ cups sugar
1¼ teaspoons ground cinnamon
1 teaspoon ground mace
1 drop clove oil

See Tips for Making Jellied Spreads (pp. 138-139). Wash and crush grapes. Separate hulls from pulp and save hulls. Heat pulp with juice and put through a colander to remove seeds. Grind hulls in a food processor, using a fine blade. Combine pulp, juice, and hulls and weigh. Place 2½ pounds of grape mixture in a large, thick-bottom saucepan. Cook on medium heat until hulls are tender. Add sugar, cinnamon, mace, and clove oil. Cook over medium heat, stirring frequently, until butter is thick and has taken on a glossiness or sheen. See Testing Fruit Butter for Doneness (p. 156). Fill hot jars immediately with the butter, leaving a ¼-inch headspace. Remove air bubbles. Wipe jar rims and adjust lids.

Process in a boiling water-bath canner.
Half-pints, 5 minutes
Pints, 10 minutes

PEACH BUTTER
Makes about 8 half-pints.

18 medium-size peaches
4 cups sugar

Use fully ripe peaches that have no insect or disease damage. See Tips for Making Jellied Spreads (pp. 138-139). Wash peaches and dip in boiling water for 30 to 60 seconds until skins loosen. Dip quickly in cold water and slip off skins. Cut peaches in half, remove pits, and chop. Place in a large, thick-bottom saucepan and cook until soft, adding only enough water to prevent sticking. Press peaches through a sieve or food mill. Measure pulp. Place 2 quarts peach pulp in a saucepan. Add sugar and cook until butter is thick and has taken on a glossiness or sheen. As mixture thickens, stir frequently to prevent sticking. See Testing Fruit Butter for Doneness (p. 156). Fill hot jars immediately with the butter, leaving a ¼-inch headspace. Remove air bubbles. Wipe jar rims and adjust lids.

Process in a boiling water-bath canner.
Half-pints, 5 minutes
Pints, 10 minutes

PEACH PINEAPPLE BUTTER
Makes about 10 half-pints.

5 pounds peaches
2 cups drained unsweetened crushed pineapple
¼ cup bottled lemon juice
2 cups sugar, or to taste

Use firm, ripe peaches that have no insect or disease damage. See Tips for Making Jellied Spreads (pp. 138-139). Wash peaches and dip in boiling water for 30 to 60 seconds until skins loosen. Dip quickly in cold water and slip off skins. Cut peaches in half and remove pits. Grind peaches in a food processor, using a medium or coarse blade. Or, crush them with a fork or potato masher. Do not use a blender. Place pulp in a large, thick-bottom saucepan and heat over medium heat. Stir constantly until peaches are tender. Place pulp in a jelly bag or a strainer lined with four layers of cheesecloth. Allow juice to drip about 15 minutes. Save the juice for later use. Measure 4 cups of the drained pulp into saucepan. Add the pineapple, lemon juice, and sugar and mix well. Heat to a boil over medium heat and boil gently about 10 to 15 minutes, stirring enough to prevent sticking. Cook until butter is thick and has taken on a glossiness or sheen. See Testing Fruit Butter for Doneness (p. 156). Fill hot jars immediately with the butter, leaving a ¼-inch headspace. Remove air bubbles. Wipe jar rims and adjust lids.

Process in a boiling water-bath canner.

Half-pints, 15 minutes
Pints, 20 minutes

Note: Non-nutritive sweeteners may be used. If aspartame is used, its sweetening power may be lost within 3 to 4 weeks.

To Vary: Use any combination of peaches, nectarines, apricots, and plums.

PEAR BUTTER
Makes about 8 half-pints.

20 medium-size pears
4 cups sugar
1 teaspoon grated orange rind
⅓ cup orange juice
½ teaspoon ground nutmeg

Use fully ripe pears. See Tips for Making Jellied Spreads (pp. 138-139). Wash pears, cut them in quarters, and remove cores. Place pear quarters in a large, thick-bottom saucepan and cook until soft, adding only enough water to prevent sticking. Press pears through a sieve or food mill. Measure pulp. Place 2 quarts pear pulp in a saucepan. Add sugar, orange rind, orange juice, and nutmeg. Heat to a boil over medium heat and boil gently about 15 minutes, stirring to dissolve sugar. As mixture thickens, stir frequently to prevent sticking. Cook until butter is thick and has taken on a glossiness or sheen. See Testing Fruit

Butter for Doneness (p. 156). Fill hot jars immediately with the butter, leaving a ¼-inch headspace. Remove air bubbles. Wipe jar rims and adjust lids.

Process in a boiling water-bath canner.
Half-pints, 5 minutes
Pints, 10 minutes

PERSIMMON BUTTER

See Tips for Making Jellied Spreads (pp. 138-139). Wash and cook persimmons in a small amount of water until soft. Strain to remove seeds and skins. Measure pulp into a thick-bottom saucepan. Add ¼ to ½ cup sugar for each cup of pulp. Spices can be added—cinnamon, cloves, and/or allspice (ground or whole: ground spices tend to make the fruit dark; whole spices, tied in a cloth bag, can be removed and fruit will remain a light color). Add enough water to prevent fruit from scorching. The grated rind and juice of one lemon or one orange can be added to each quart of pulp, if desired. Heat to a boil on medium heat and boil gently, stirring often to prevent scorching. Cook until butter is thick and has taken on a glossiness or sheen. See Testing Fruit Butter For Doneness (p. 156). Fill hot jars immediately with the butter, leaving a ¼-inch headspace. Remove air bubbles. Wipe jar mouth and adjust lids.

Process in a boiling water-bath canner.
Half-pints, 5 minutes
Pints, 10 minutes

PLUM BUTTER
Makes about 8 half-pints.

4 pounds plums	½ teaspoon ground cinnamon
4 cups sugar	½ teaspoon ground cloves
⅓ cup lemon or orange juice	½ teaspoon ground allspice

See Tips for Making Jellied Spreads (pp. 138-139). Wash plums; crush them and save the juice. Cook plums and juice until skins are soft, stirring constantly. Put plums through a fine sieve to remove skins and seeds. Prepare enough to make 2 quarts of pulp. Place pulp in a large, thick-bottom saucepan. Add sugar and bring to a boil over medium heat, stirring to dissolve sugar. Then, add juice, cinnamon, cloves, and allspice. Increase heat and boil rapidly, stirring constantly to prevent scorching. As cooking progresses, reduce heat somewhat to prevent spattering. Cook until butter is thick and has taken on a glossiness or sheen. See Testing Fruit Butter for Doneness (p. 156). Fill hot jars immediately with the butter, leaving a ¼-inch headspace. Remove air bubbles. Wipe jar rims and adjust lids.

Process in a boiling water-bath canner.
Half-pints, 5 minutes
Pints and Quarts, 10 minutes

QUINCE BUTTER

See Tips for Making Jellied Spreads (pp. 138-139). Wash quince and cut into small pieces. Measure it into a large, thick-bottom saucepan. Add ½ cup water for each cup of quince. Cook over medium heat until fruit is soft, stirring frequently. Put cooked fruit through a fine sieve. Measure pulp into saucepan. Add ½ cup sugar for each cup of pulp. Bring to a boil over medium heat, stirring to dissolve sugar. Then, boil rapidly, stirring constantly to prevent scorching. As cooking progresses, reduce heat somewhat to prevent spattering. Cook until butter is thick and has taken on a glossiness or sheen. See Testing Fruit Butter for Doneness (p. 156). Fill hot jars immediately with the butter, leaving a ¼-inch headspace. Remove air bubbles. Wipe jar rims and adjust lids.

Process in a boiling water-bath canner.
Half-pints, 5 minutes
Pints, 10 minutes

FRUIT SYRUP RECIPE

Syrups made from fresh or frozen blueberries, cherries, grapes, raspberries, and strawberries make excellent toppings for pancakes, fruits, and desserts.

BERRY SYRUP
Makes about 9 half-pints.

6½ cups fresh or frozen berries   6¾ cups sugar

Use freshly picked, ripe, sound berries. Or, you can use frozen berries. See Tips for Making Jellied Spreads (pp. 138-139). Wash fresh berries and remove caps and stems. Crush berries in a large, thick-bottom saucepan. Heat to boiling, reduce heat, and simmer until soft, about 5 to 10 minutes. Strain hot berries through a sieve and drain until cool enough to handle. Strain the collected juice through a double layer of cheesecloth or jelly bag. Discard the dry pulp. Measure 4½ to 5 cups juice into saucepan. Add sugar and bring to a boil, stirring to dissolve sugar. Reduce heat and simmer 1 minute. Remove from heat and quickly skim off foam. Fill hot jars immediately with syrup, leaving a ½-inch headspace. Wipe jar rims and adjust lids.

Process in a boiling water-bath canner.

Half-pints or Pints, 10 minutes

To Vary: For a syrup containing whole pieces of berries, set aside 1 or 2 cups of berries. Then, add them when you add the sugar.

REFRIGERATED SPREAD RECIPES

The following jellied fruit spreads are tasty, yet lower in sugar and calories than regular jellies and jams. Gelatin is used as a thickening agent.

REFRIGERATED APPLE SPREAD
Makes 4 half-pints.

1 quart unsweetened apple juice
2 tablespoons bottled lemon juice
2 tablespoons unflavored gelatin
2 tablespoons liquid low-calorie sweetener
Food coloring, if desired

Combine apple and lemon juices in a saucepan. Sprinkle the gelatin over the juices to soften. To dissolve gelatin, heat juice to a full rolling boil and boil 2 minutes. Then, remove from heat and stir in sweetener and food coloring. Fill hot jars immediately with spread, leaving a ¼-inch headspace. Wipe jar rims and adjust lids, but do not process in a canner and do not freeze. When jars have cooled, store them in the refrigerator.

To Vary: For spiced apple jelly, tie 2 crushed cinnamon sticks and 4 whole cloves in a spice bag. Add bag to mixture before boiling. Remove spice bag before adding the sweetener and food coloring.

Caution: Use within 4 weeks.

REFRIGERATED GRAPE SPREAD
Makes 3 half-pints.

1 24-ounce bottle unsweetened grape juice
2 tablespoons bottled lemon juice
2 tablespoons unflavored gelatin
2 tablespoons liquid low-calorie sweetener

In a saucepan, combine the grape and lemon juices. Sprinkle the gelatin over the juices to soften. To dissolve gelatin, heat juices to a full rolling boil and boil 1 minute. Then, remove from heat and stir in sweetener. Fill hot jars immediately with spread, leaving a ¼-inch headspace. Wipe jar rims and adjust lids, but do not process in a canner and do not freeze. When jars have cooled, store them in the refrigerator.

Caution: Use within 4 weeks.

Canning for Fairs and Exhibits

Many Alabama homemakers like to show their home-canned products in county and state fairs. A lot of skill and know-how goes into blue-ribbon products, and the winners can be justly proud.

For award-winning home canned products, the following suggestions may help. There are certain qualities that judges look for when they are scoring entries. Winning canners work for these qualities:

- Quality of product.
- Quality of pack.
- Quality of liquid.
- Appearance of jar.

1. Quality of Product by Appearance 45
   a. Uniform in size and shape ...... (15)
   b. Good condition; natural shape retained ...... (15)
   c. Fresh, natural color ...... (15)
2. Quality of Pack 15
   Full jar with good proportion of product to juice.
3. Quality of Liquid 30
   a. Clear and free of sediment ...... (15)
   b. Clear, natural color ...... (15)
4. Appearance of Jar 10
   a. Clean, standard canning jar ...... (5)
   b. Label of appropriate size and color, neatly placed and clearly marked ...... (2)
   c. Shiny, new lids and bands ...... (3)

Total 100

HOW TO WIN AT CANNING

Each type of food has its own characteristics to consider in preparing a 1st place product. Generally, jars will not be opened nor product tasted by the judges. So quality is judged by appearance. However, this is not always the case. Flavor and texture are sometimes considered.

In most cases, you need to be careful in **selecting, preparing, packing, and processing** food to be a winner.

General Guidelines

• All entries must have been canned by the person making the entry.

• All food entered in an exhibit should have been canned within the last food preservation season.

• All vegetables (not tomatoes) and meats should have been canned at 10 pounds pressure in a pressure canner.

• All fruits, tomatoes, juices, jams, jellies, preserves, pickles, relishes, marmalades, butters, and conserves should have been processed in a boiling water-bath canner.

• All entries should be in clean standard canning jars in good condition with new, 2-piece lids (self-sealing lid and band). Jars and lids should be the same brand. If two or more jars are entered, they should be the same brand, size, and shape.

• All jars should be neatly packed. The product should be level at the top and covered by liquid, if needed. The recommended amount of headspace should be left above the liquid.

• All jars should be neatly labeled with the name of the product and canning date.

• There should be no signs of spoilage, such as gas bubbles, leaky seals, bad odor, or cloudy liquid. Jars should be sealed when tested by judges.

Vegetables and Fruits

Selecting. Vegetables and fruits at their prime stage of maturity have the best texture, color, flavor, and nutritive value. Do not use any that are over or under mature (ripe). They should be tender and firm—not mushy, tough, or stringy. The color should be natural and uniform. Never use artificial color. Select pieces of the same shape and size to pack together. And, of course, no food should have spots or blemishes.

Preparing. The natural shape of the food should be preserved, if possible. Leave small vegetables and fruits whole. Those that are too large should be cut into a uniform shape and size. Some can be halved. Tomatoes should be quartered for an economy pack. Cut snap beans smoothly rather than snapping them. Sprinkle ascorbic acid mixture over fruits to keep them from turning dark.

If precooking vegetables or fruits, cook just long enough to look cooked but not overcooked. After heating fruits, let them sit immersed in the syrup until they cool. The fruits absorb the syrup and then won't float as easily in the jars.

Packing. Pack the food well—not fancy. The space should be filled but not crowded. Leave space for enough liquid to allow proper heat penetration during processing. This is important for safety and for good flavor. Greens and starchy vegetables, such as corn, shelled beans, and peas, should have enough liquid to make a loose pack when processed.

Fill non-starchy vegetables and fruits to within ½ inch of the top of jar. Fill starchy vegetables, such as corn, shelled beans, and peas to within 1 inch of the top. Pack jars neatly and attractively. For example, pack peach halves with blossom end toward outside of jar. Add liquid as you fill the jar.

Liquid for Vegetables. Liquid should be clear and free from cloudiness and sediments. Liquid in green shelled blackeye and field peas, however, may be slightly cloudy. The liquid should cover the food but should come no higher than ½ inch from the top of the jar.

Discard the water used to precook the vegetables. Fresh water makes a clearer liquid. Use soft water and do not add salt. Salt may cause a cloudy liquid.

To make hard water soft, boil it for 15 minutes in a stainless steel or uncracked enamel boiler. Then, cover it and let it sit for 24 hours. Next day, slowly pour the water into a clean jar, being sure not to disturb the minerals that have settled to the bottom.

Liquid for Fruits. Liquid should be a syrup. Make the syrup according to a recipe for the fruit you are canning. The syrup should be free of any sediment. Make a fresh syrup or strain the syrup the fruit was precooked in through a cloth. The syrup should cover the fruit but should come no higher than ½ inch from the top of the jar.

Quartered tomatoes will produce enough juice of their own. Add tomato juice to whole tomatoes.

Use a **plastic** spatula to remove all air bubbles after packing jars.

Meats

Selecting. Fully mature meat is best for canning. Meat that is too young is easily ruined by the high temperature of a pressure canner. However, for best eating, the meat should not be too old, tough, or stringy. The texture should be reasonably fine.

Preparing. Cut meats into serving-size pieces. The pieces in each jar should be as uniform in size and shape as possible. It is not necessary to cut away all fat, but there should not be too much. A line not more than ¼ to ½ inch deep is acceptable. Too much fat slows heat penetration during processing.

Meat can be packed into jars either raw or precooked. If precooked, it should look moist and full of flavor, not over-cooked, hard, and dry. The color should be characteristic of the meat when it is heated or cooked. If it is to be browned, brown it only slightly. Too much browning gives the meat an overcooked flavor.

Packing. Pack meat neatly and attractively to within 1 inch of the top of the jar. Pack it snugly, particularly next to the sides of the jar, but not tightly. It should be looser through the center of the pack for the liquid to allow good heat penetration during processing.

Liquid for Meats. Liquid should be the precooking liquid or the natural meat juices. For a hot pack, the liquid should come to the top of the meat. It should be fairly clear, preferably jellied. For a raw pack, do not add liquid. Fried meats may be canned with or without liquid.

Jellied Products

At some fairs, jellied products—jelly, jam, preserves, marmalade, and fruit butter—are opened for judging. This makes characteristics other than appearance more important.

Usually, pint or half-pint jars with 2-piece lids are best for jellied products.

General Characteristics. The **products** should be like the fruits they are made from. They should have a full, fresh-fruit flavor. They should not be too tart or too sweet. They should not have an off-flavor of any kind. The color should be the same as the fruits used. A dark color indicates overcooking or scorching.

Special Characteristics. Special characteristics do not apply to all jellied products but only to the types specified.

Jelly. Jellies should be clear and sparkling. When turned from the jar, they should quiver, cut easily with a spoon, and retain their shape when cut. They should be very tender when cut—not tough, stringy, thin, or syrupy. There should be no crystals.

Preserves and Marmalade. The pieces of fruit in preserves and marmalades should be cut into uniform sizes and shapes. They should retain their shape during precooking. They should become thoroughly saturated with a heavy syrup, leaving them transparent and plump, not shriveled. The fruit should be tender when cut. The syrup should be thick enough to move slowly with a definite pull when the jar is turned to one side. The pull and thickness will differ with the kind of fruit used. The color of the syrup should be the same as the fruit, clear and shiny.

Fruit Butter. Butters should move very slowly with a strong pull from the side of the jar when turned to one side. It may move in a solid mass.

Pickles

Pickles, like jellied products, are more likely to be opened at fairs than other canned products. This makes texture and flavor more important for judging.

Texture. Pickles should be firm, yet tender. They should be crisp and plump, never watery. They should be cut into uniform sizes and shapes and should hold their original shape.

Flavor. Pickles should have the natural flavor of the food or combination of foods from which they are made. The pickles should not be too sour, too sweet, or too spicy. They should be thoroughly saturated with the pickling solution or syrup.

Color. Pickles should be the natural color of the foods from which they are made. Cured and fresh-pack cucumbers should change from a bright green to olive green. **Never use artificial coloring.** In many cases, pickles should look transparent or semi-transparent.

Freezing

Freezing is a quick, convenient, and easy method of preserving foods at home. In many cases, frozen foods are ready to serve on short notice because most of the preparation is done before freezing. Freezing preserves nutritive quality as near to that of fresh foods as any food preservation method used.

FOOD SAFETY AND QUALITY

Food properly selected, prepared, packaged, frozen, and stored will keep its natural color, flavor, texture, and nutritive value, and it will be safe to eat. However, no matter how carefully products are prepared and processed, if the quality is poor before freezing, it will remain poor.

Loss of either food safety or quality during freezing may be caused by: (1) the chemical action of enzymes; (2) ice formation during freezing; (3) surface drying or "freezer burn"; (4) microorganisms; or (5) unfavorable storage conditions or storing too long. (See Tables 14-16, pp. 179, 180, and 182.)

Enzymes

All foods contain enzymes which cause both physical and chemical changes. Some of these changes are desirable. Beef, for example, is aged in a chill room for 5 to 10 days to give the enzymes a chance to make the meat tender. Likewise, enzymes in plant tissues cause the fruit or vegetable to mature and reach its peak of quality. On the other hand, enzymes also cause plant foods to discolor and decay.

Enzyme action can be controlled by heating or freezing. Freezing slows enzyme activity so that many foods, including meats and fruits, will keep satisfactorily with little or no further control. Surface browning of a cut fruit, however, is an enzyme reaction that can take place before the fruit can be frozen. This reaction can be controlled by coating the fruit with a syrup or dipping it into an ascorbic acid solution as soon as it is cut.

Heating is required to stop the enzyme activity in vegetables. Heat destroys enzymes and is used to prevent changes in the food's color, flavor, and texture.

Ice Formation

When food is put in a freezer, the moisture in the food freezes and forms ice crystals. Quick freezing changes the moisture to very small ice crystals. Small crystals are desirable because they do not tear or damage the food tissues. Slow freezing results in large ice crystals which tear food tissues and cause the food to appear limp or soft after thawing. Food should be frozen quickly and stored at a constant temperature of 0°F or lower.

Freezer Burn

The air in most freezers is very dry. Cold air drops from the coils toward the floor, circulating around the frozen food and soaking up moisture. Through weeks of storage, this dry air removes the ice from any foods exposed to air, often causing a dry, tough, discolored surface to develop. This is called "freezer burn." To prevent it, always use a moisture-vapor-resistant wrap or container and make sure the food is packaged and sealed airtight.

Microorganisms

Bacteria, molds, and yeasts are microorganisms that are present in all foods. They multiply rapidly when the temperature is between 40° and 140°F. Freezing does not kill most microorganisms in food, but it does prevent their growth if the food is held at 0°F or below. When thawed, the surviving organisms can grow again; thus, proper handling and preparation at that time are essential.

Storage Conditions

Under the best conditions, most uncooked foods will keep their quality in a freezer for 6 months to a year. The best conditions include being packaged airtight in a moisture-vapor-resistant container or wrap and held in a properly designed freezer at a continuous 0°F or less. Under poor conditions, off-flavor and -color may develop within 3 months or less. When freezer temperature fluctuates, foods can partially thaw and then refreeze. This often results in large ice crystal formation and a poor quality product. (See Table 15, p. 180, for storage times.)

GENERAL STEPS IN FREEZING

1. If you are preparing to freeze a large quantity of food, set your freezer's temperature control to -10°F 24 hours before you will be putting the food in the freezer (see Freezers, pp. 173-175).

2. Select the varieties of vegetables or fruits that freeze well. Use only fresh, tender yet mature, good quality vegetables and fruits; use healthy, good quality animals. Foods will not improve while frozen; start with foods that are the quality you would choose when preparing a meal (see Food Safety and Quality, pp. 168-169).

3. Wash vegetables and fruits thoroughly until every trace of soil or chemical residue is removed. Sort for size. This is especially important for vegetables which must be blanched so that all pieces will be treated evenly. Dress and chill meat, poultry, and seafood (see instructions for preparing specific foods: fruits, pp. 183-186; vegetables, pp. 194-196; meats, pp. 203-205; poultry, pp. 205-206; seafood, p. 206; eggs, p. 209; nuts, p. 210; and dairy products, p. 211).

4. Assemble and thoroughly wash freezer containers (see Packaging Materials, pp. 171-172).

5. Get food ready for packaging (see recipes or instructions: fruits, pp. 186-193; vegetables, pp. 197-203; meats, pp. 203-205; poultry, pp. 205-206; seafood, pp. 207-208; eggs, p. 209; nuts, p. 210; and dairy products, p. 211). Cut it into pieces that are the best size for the way it will be used. Work quickly to keep food quality from deteriorating.

6. Prepare syrup (see Table 17, p. 185), broth, or sauce that may be used in freezing.

7. Treat fruits to prevent darkening (see Pretreating, pp. 183-184); blanch vegetables (see Blanching, pp. 194-195).

8. Package serving-size portions of prepared food in moisture-vapor-resistant packaging materials (see Packaging Food, pp. 175-179). Securely fasten lids on rigid containers and seals on flexible bags.

9. Label containers with name of food, number of servings or amount, date it was put in the freezer, and any special information (see Keeping Records, p. 179).

MATERIALS AND EQUIPMENT

Needed:
- Freezer
- Containers (see Packaging Materials, pp. 171-172)
- Blanching equipment (see Blanchers, p. 175)
- Timer or clock
- Colander or sieve
- Measuring cup
- Sharp knife
- Clean cloths
- Hot pads

Helpful:
- Quart measuring cup
- Potato masher or pastry blender
- Kitchen shears

Packaging Materials

An important part in frozen food storage is the container or wrap the food is packed in. Good materials can prevent freezer burn and breakdown of texture in frozen foods; they also prevent color and flavor changes. Good materials for freezing have the following characteristics:

- They are airtight when properly closed or sealed; they will protect the food from absorbing off-flavors or odors.
- They are moisture-vapor-resistant so that no liquid, vapor, nor air can get in or out of the packaged product. Most packaging materials for frozen foods are not moisture-vapor-proof, but are sufficiently moisture-vapor-resistant to retain satisfactory quality of frozen foods during storage.
- They are durable and able to withstand low temperatures; they will not become brittle or crack.
- They will not absorb oil, grease, and water. They can be easily cleaned when the food is removed.
- They are odorless and tasteless.
- They are easy to stack and take as little space in the freezer as possible.
- They are easy to handle, easy to seal, and easy to label.

There are two types of packaging materials for home use: rigid containers and flexible bags or wrappings.

Rigid Containers. Most foods can be stored in rigid containers. They stack well and can be reused. They are especially good for foods that contain liquids. In most cases, foods packed in liquids should not be frozen in glass jars. Round containers generally take more freezer space.

The most popular rigid container is made of moisture-vapor-resistant plastic with a snap-on lid. With proper care, plastic containers can be used for years. Wash lids in warm, never hot, water. Be careful as you remove the lid from the container because the lid may stretch and not fit securely. Then, the container will not be airtight. It is best to let cold water run over the lid before removing. If the lid stretches, use freezer tape to seal it securely on the container.

Glass, aluminum, tin, or heavily waxed cardboard containers are suitable if they have tight-fitting lids as well as the characteristics listed above.

If using glass jars, be sure to choose the wide-mouth, dual-purpose jars that are for freezing and canning. These jars are specially made to withstand freezing and boiling temperatures. The wide mouth allows removal of partially thawed foods and gives room for expansion during freezing. Allow at least 1 inch headspace when freezing in glass. **Use new lids** each time and rinse them in cold tap water before applying to the jar mouth.

Foods can be frozen in standard canning jars that have shoulders, but there is danger of breakage if the jar is filled above the shoulder. Food must be thawed before removing from these jars. Extra care should be taken when thawing food in glass.

Caution: Do not use glass jars to freeze liquids.

Flexible Bags. Bags made of moisture-vapor-resistant polyethylene are suitable for any frozen foods packed dry. They can also be used for liquid packs, but they hold up better if used with a cover box.

When sealing a freezer bag, press out any air that might be trapped in the bag, twist the top of the bag to form a spiral, and fold it over like a gooseneck. Wrap the gooseneck several times about ½ inch from the food with a rubber band, plastic-covered wire, or some similar closure material. If a covered-wire closure is used, be sure to wrap the ends so they won't puncture the bag. The bag should be loose to allow food to expand during freezing.

Collapsible Cover Boxes. Used with a moisture-vapor-resistant freezer bag, a cover box is a most economical container. It protects the bag from getting punctured, and it makes stacking the packages easier. With good care, cover boxes can be reused for several years, but it is better to use new freezer bags each time.

Freezer Wrap. Freezer paper is a good material for wrapping meats and certain other large or irregularly shaped foods to be frozen; freezer cellophane, freezer aluminum foil (0.0015 thickness), and pliofilm may also be used. If freezer paper is selected, choose only the type made especially for this purpose. Do not confuse ordinary wrapping cellophane or butcher paper with specially developed grades recommended for freezing. Household wax paper and plastic wrap are not recommended; neither is most heavy-duty aluminum foil. Extra-heavy-duty aluminum foil can be used for freezer storage for up to 8 weeks. Meats bought at the supermarket should be wrapped again in suitable freezer wrap before freezing. The supermarket wrap is porous and may contain small holes.

Freezer Tape. Using freezer tape to seal packages of food to be frozen makes the packages more airtight. Such tape must be able to hold a tight seal at 0°F or below.

Unsuitable Packaging Materials. The following containers are not moisture-vapor-resistant enough to use for long-term freezer storage. The food will probably develop freezer burn and may absorb undesirable odors or flavors.

- Wax paper
- Paper cartons
- Cottage cheese cartons
- Cardboard ice cream or milk cartons
- Any rigid container that may crack
- Any container that has a poorly fitting lid

Freezers

Home freezers should maintain a constant temperature of 0°F or lower and should be able to freeze food quickly. Low temperature is necessary for fast freezing, and fast freezing is necessary for a good frozen product. Place a freezer thermometer in the center of the freezer to make sure it is freezing properly (0°F or lower).

The more rapidly food is frozen, the better quality it will be. For this reason, it is best to have a whole beef side or pork frozen at a reliable local freezer plant. However, if you must depend on your home freezer, set the temperature control at -10°F or lower about 24 hours before you will add the unfrozen food. Be careful not to put too much unfrozen food in the freezer at one time because the product may sour before freezing. Or, it may cause undue warming of previously frozen products.

Never add within a 24-hour period more than 2 to 3 pounds (pints) of unfrozen food per cubic foot of freezer space. In other words, if you have a 12-cubic-foot freezer, then 24 to 36 pounds (pints) are as many as you should place in the freezer until that food is completely frozen, about 12 to 24 hours later. Place each package of unfrozen food in the coldest part of the freezer. Allow a little space around each container until packages are frozen; then pack them close together. Keep frozen at 0°F or below.

Refer to the manufacturer's manual that came with your freezer for information on operating the freezer.

Storage Costs. Operating a freezer can be costly. For economical operation of your home freezer, keep it filled. About 35 pounds of food can be stored in one cubic foot of freezer space. When you use food from your freezer, replace it as soon as possible. A half-empty freezer is a costly freezer. Also, a freezer does not operate efficiently if ice builds up on the inside walls. If your freezer does not defrost automatically, you will need to defrost it periodically (see Freezer Care, below).

To get the most from a rental freezer-locker, keep it full. The rent is the same whether the locker is full or only half-full.

Freezer Care. Carefully read the use-care booklet that came with your freezer and follow the manufacturer's recommendations for care. If your freezer is not a frost-free model, defrost it once a year or when ¼ to ½ inch of frost covers a large area of the sides. Most people prefer to defrost and clean the freezer in the spring before a new supply of fruits and vegetables is added. Frost-free freezers do not need to be defrosted but do need to be cleaned once a year or more often if it is visibly soiled. If you cannot find your use-care booklet, follow these general recommendations.

Cleaning. Remove the food to insulated ice chests or large cardboard boxes. Pack boxes tightly and cover with newspapers or a quilt for insulation. Unplug the freezer. Wash the inside of the freezer, using a solution of 1 tablespoon baking soda to 1 quart water. Do not use soap. Rinse and wipe the freezer dry, leaving it open long enough to air. Work quickly to get the freezer running again before food begins to thaw. Plug in the freezer and reset it to 0°F. When temperature in freezer is near 0°F, put packages back in the freezer.

Defrosting. Unplug the cord and remove the food. Pack the food the same as for Cleaning (p. 173). Scrape frost from freezer sides with a dull-edged plastic scraper or wooden spoon. Be careful not to puncture or otherwise injure the inside walls of the freezer. Do not use a knife or other sharp tool to scrape sides. Scoop frost out of the bottom of the freezer and wipe up any remaining moisture. For quicker defrosting, pour cold water down the inside walls of the freezer. Have clean towels ready to absorb the water. Then, wash the inside of the freezer, reset freezer, and return food (see Cleaning, above).

Removing Odors. To eliminate undesirable odors, unplug the freezer and remove the food. Pack the food the same as for Cleaning (p. 173). Wash the inside of the freezer with a solution of 1 cup vinegar to 1 gallon water. Allow it to dry without rinsing. If the odor remains, arrange to store your frozen food for several days. Then, place activated charcoal on paper in the bottom of the freezer. Be sure to use activated charcoal, not the cooking type. You can buy it at pet supply stores or drug stores. Close the freezer door and let it sit for several days. You may need to replace the charcoal after a few days. When the odor is gone, rinse and dry the inside of the freezer. Plug it in and allow the temperature to reach near 0°F. Then, replace the food.

Power Failure. In case the electric current goes off, food will usually stay frozen in a fully loaded freezer for two days, **if the freezer was operating at 0°F or below at the time of the power failure.** In a freezer that is only half-full, the food will keep only one day. Do not open the freezer door any more than necessary.

Dry ice can be placed in the freezer to keep the temperature down. Use about 2 pounds of dry ice per cubic foot of freezer space. Use gloves to handle dry ice. Place cardboard over the food packages before putting dry ice in the freezer.

Some freezer-locker plants may have available space you can rent in an emergency. Call and find out before taking your food out of your freezer. In moving frozen food to another freezer, pack it as tightly as possible in cardboard boxes and fill empty spaces with newspaper. Cover the boxes with paper and blankets.

You may safely refreeze foods that have thawed **if they still contain ice crystals, are still cold (40°F), and have been held at this temperature no longer than 1 or 2 days after thawing** (see Refreezing Food, pp. 179-181). Foods warmed to 40°F or higher are not recommended for refreezing. Remember, any food that has partially thawed loses color, flavor, and texture.

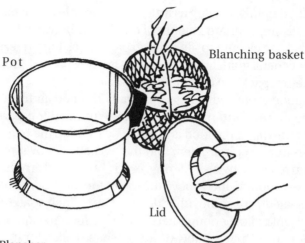

Figure 12. Water Blancher

Fruits usually ferment when they start to spoil. A little fermentation will not make fruits dangerous to eat, but it may spoil their flavor. So, you can refreeze thawed fruits if they taste and smell good. Or, you can use them in cooking and baking or for making jams, jellies, and preserves.

Blanchers

Almost all vegetables need to be blanched before freezing to stop enzyme action that affects the color, flavor, and nutritive value of the food. Food can be blanched in boiling water or in steam.

Water Blanchers. A water blancher is basically a large pot with a tight-fitting lid and a wire basket that fits inside to hold the vegetables in boiling water for a short time. You can buy a standard vegetable blancher or use a metal colander, sieve, deep-fryer basket, or a cheesecloth bag that will fit inside a large pot with a tight-fitting lid.

Steamers. Some vegetables and a few fruits need to be steam blanched. You can buy a standard food steamer or use any of the equipment listed above for water blanching as long as the basket or a metal rack will hold the food 3 inches above 1 to 2 inches of rapidly boiling water.

METHODS AND PROCEDURES

Packaging Food

• Cool all foods and liquids before packing. This will speed freezing and help foods retain their natural color, flavor, and texture.

• Pack food in quantities that will be used for a single meal.

• Follow the recipe for freezing each food to determine whether it should be packed dry or in liquid.

176 • Freezing/General

- Pack food tightly, leaving as little air as possible in the package.
- Leave the appropriate amount of headspace between the top of the food and the package closure to allow for expansion as food freezes. See Table 14 (p. 179) for general guidelines. Foods that are tray packed (frozen individually on trays before packing) do not need headspace.
- When food is packed in bags, press any trapped air from the bag, beginning at the bottom and moving toward the unfilled top part. Press the bag together at the top of the food and seal it by twisting the bag above the top of the food to form a spiral. Then, fold the twisted part over to form a "gooseneck". Wrap a plastic twist-tie, string, or a rubber band around the gooseneck several times about ½ inch above the top of the food. Some bags may be heat sealed or sealed with some other device. The ½-inch space above the top of the food should allow enough headspace for food to expand. Some recipes specify more headspace. Allow whatever space called for in the recipe.
- Fasten lids on rigid containers carefully. Use a tight lid and keep the sealing edge free of moisture or food to ensure a good closure. Secure loose-fitting lids with freezer tape.
- Wrap meats, breads, and other large or irregularly shaped foods in freezer wrap, using either the drugstore wrap (p. 177) or the butcher wrap (p. 178).
- Label each package (see Keeping Records, p. 179), using freezer tape, marking pens or crayons, or gummed labels that are made specifically for freezer use.

Using the Drugstore Wrap

This is the preferred wrap method for meat, fish, and poultry parts (Figure 13). It is also appropriate for food products such as sandwiches and some other prepared foods.

1. Cut a piece of freezer paper large enough to go completely around the food plus 5 to 6 inches. Place the food in the center of the paper. If you are wrapping several pieces of meat in one package, place two pieces of wax paper between each piece to prevent them from freezing together. Pad bones with extra pieces of freezer paper so they won't puncture the outside wrapping.
2. Raise the edges of two opposite sides and bring them together over the food.
3. Fold the raised edges over about ½ to ¾ inch.
4. Continue folding toward the food until the paper fits snugly across the top. Smooth the fold to fit neatly across the package.
5. Turn the package over and press against each end of the food to remove air.
6. Fold the two ends to make triangles, pressing out air pockets.
7. Bring the ends up and over the package to the center. Tape the ends down with freezer tape. You may want to overwrap packaged meat with heavy paper to protect the freezer paper from getting torn. Moving frozen food around in the freezer may tear the freezer paper.
8. Label and date the package.

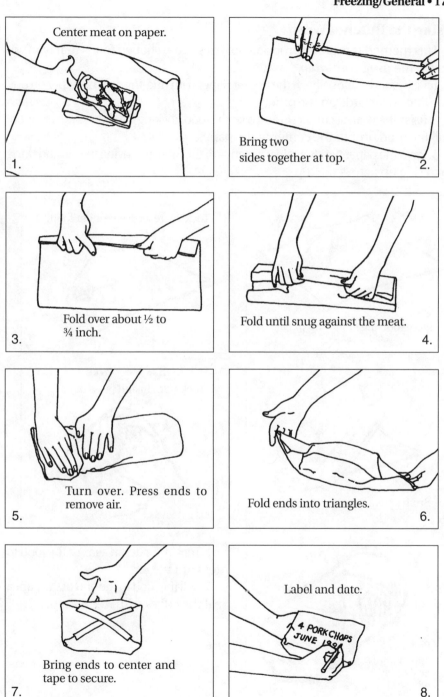

Figure 13. Drugstore Wrap

Using the Butcher Wrap

This method is more appropriate for irregularly shaped foods that won't fit well in the drugstore wrap.

1. Place food diagonally on the freezer paper with one-third of the paper on one side and two-thirds on the other.

2. Raise the nearest corner up and over the food, tucking it tightly under the food. Roll the food over to the center of the paper.

3. Press the paper tightly against the ends of the food, removing the air, and bring each side up across the top.

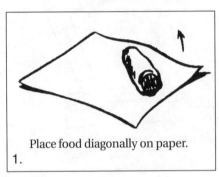

Place food diagonally on paper.
1.

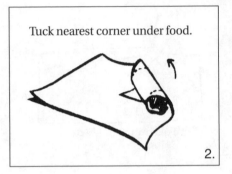

Tuck nearest corner under food.
2.

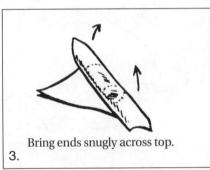

Bring ends snugly across top.
3.

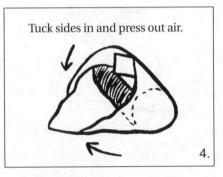

Tuck sides in and press out air.
4.

4. Tuck the sides in against the food to press out the air.

5. Roll the food to the end of the paper. Seal the edges with freezer tape.

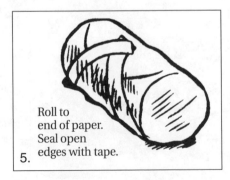

Roll to end of paper. Seal open edges with tape.
5.

Figure 14. Butcher Wrap

Table 14. Headspace to Allow in Freezer Packages

Type of Pack	Freezer Bags and Wide-Top Containers		Narrow-Top Containers	
	Pint	Quart	Pint	Quart
Liquid pack*	½ inch	1 inch	¾ inch**	1½ inches
Dry pack***	½ inch	½ inch	½ inch	½ inch

*Fruits packed in any liquid or sugar, crushed or pureed fruits, or juices.
**Headspace for juices should be 1½ inches.
***Fruits or vegetables packed without added sugar or liquid.

Keeping Records

Recording what you put in and take out of the freezer will help you make the best use of your frozen foods. If kept near your freezer, such a list can be easily kept up to date. Individual packages should also be labeled with the name of the food, any added ingredients, the number of servings or amount, the way the food was prepared, and the date it was placed in the freezer.

Sometimes, packages can get "lost" in the back or bottom of the freezer and won't be used before they become too old. A good record will let you know exactly what foods you have and how long they have been in storage. Include the name of the food, how it was prepared, the number of servings or amount per package, the number of packages, and when they were frozen. Allow an easy way to mark removal of each package. All frozen cooked foods should be used in about 3 months or less. See Table 15 (p. 180) for recommended storage times for other home-frozen foods.

To save time and work, plan a system of storing foods in your freezer. For example, store all vegetables in one section, all meats in another, all fruits in a third, and cooked foods in the fourth. In the vegetable section, you can put all corn in one row, all peas in another, etc. This way, you'll know where to look for what you want, and you won't have to move too many packages to find the right one.

Refreezing Food

If your power goes out or your freezer malfunctions, your food may begin to thaw. In such cases, you may be able to refreeze the food, but you must be careful.

Examine each package of meat, vegetable, fruit, or cooked food before you decide what to do with it. If the color or odor of the thawed food is poor or questionable, discard it. It may be dangerous!

Ground meat, poultry, and seafood become unsafe to eat more quickly than other foods. Bacteria multiply rapidly in these foods, so don't refreeze any of them if they have thawed completely. If ice crystals are still in the food, it is usually safe to refreeze it immediately, even though the quality may suffer.

Table 15. Recommended Storage Times for Home-Frozen Foods

Food	Storage at 0°F or Below
Butter or Margarine	9 months
Cheese	
Dry-Curd Cottage, Ricotta	2 weeks
Natural, Processed	3 months
Cream	2 months
Whipped	1 month
Egg Whites or Yolks	12 months
Fish or Shellfish	
Fatty Fish	3 months
Lean Fish	6 months
Shellfish	3 months
Fruits (except Citrus)	8 - 12 months
Citrus Fruits and Juices	4 - 6 months
Ice Cream or Sherbet	1 month
Meat	
Bacon	1 month
Frankfurters	2 months
Ground or Stew Meat	3 months
Ham	2 months
Beef or Lamb Roasts	12 months
Pork or Veal Roasts	8 months
Beef Steaks or Chops	12 months
Lamb or Veal Steaks or Chops	9 months
Pork Steaks or Chops	4 months
Variety Meats	4 months
Milk, Fresh Fluid	1 - 3 months
Poultry	
Cooked, with Gravy	6 months
Cooked, no Gravy	1 month
Uncooked Whole Chicken or Turkey	1 year
Uncooked Whole Duck or Goose	6 months
Uncooked Chicken Parts	9 months
Uncooked Turkey Parts	6 months
Vegetables	8 - 12 months
Yogurt, Regular	
Plain	1 month
Flavored	5 months

Refreeze food quickly to protect its quality and safety. Turn the freezer's temperature control to its coldest position. The motor will run continuously until the food is frozen. Place the warmer packages against the refrigerated surface, if possible; place them so air can circulate around all of them. If the freezer is too full, move some of the colder packages to the refrigerator; then, return them gradually to the freezer **but only if they still contain ice crystals.** After the food is well frozen, turn the temperature control to its usual setting.

If your freezer is full of warmed foods, it is best to take them to a commercial locker plant. Chill them to 0°F or below before taking the food back to your home freezer. Wrap the food well with newspapers and blankets before moving it to or from the freezer plant.

Foods that have been frozen and thawed require the same care as foods that have never been frozen. Use refrozen foods as soon as possible.

COMMON CAUSES OF PROBLEMS

Freezer Burn

Foods become dry, tough, and discolored when:
- Packages are torn or not fastened securely.
- Packaging material is not moisture-vapor-resistant.
- Too much air is left in package.

Gummy Liquid in Fruits

Liquids are affected when:
- Fruits are frozen too slowly.
- Freezer temperature is too warm.
- Freezer temperature fluctuates during storage.

Rancid Flavor in a Product Containing Fat

Canned meats and vegetables flavored with pieces of meat or fat develop a rancid flavor if:
- Too much air is left in package.
- Vegetables are not blanched correctly.
- The package is not fastened securely.
- The package is stored too long.

Grassy Flavor in Vegetables

This undesirable flavor may develop when:
- The vegetables are not blanched.

Mushy Food

Changes in texture are often the result if:
- Too much food is put in the freezer at one time.
- Food is frozen at a temperature higher than 0°F.
- Freezer is too warm during storage.

Green Vegetables Turn Olive Brown

Color change means enzymes are still active because:
- The vegetables are not blanched.

Table 16. Foods Not Suitable for Freezing

Foods	Results of Freezing
Cabbage, Celery Cucumbers, Lettuce, Radishes, Endive	They become limp, water-logged, and quickly develop oxidized color, flavor, and aroma.
Irish Potatoes	They become soft, crumbly, water-logged, and mealy.
Cooked Pasta, Rice	They become mushy and taste warmed over.
Cooked Egg Whites	They become soft, tough, rubbery, spongy.
Meringue	It becomes tough.
Icings from Egg Whites	They become frothy and weep.
Cream/Custard Fillings	They separate; become watery and lumpy.
Milk Sauces	They may curdle or separate.
Sour Cream	It separates and becomes watery.
Cheese/Crumb Toppings	They become soggy.
Mayonnaise	It separates.
Gelatin	It weeps.
Fruit Jelly	It may soak into bread.
Fried Foods	Most lose crispness and become soggy.

Note: Cabbage and cucumbers freeze well when marinated for freezer slaw or freezer pickles, which do not have the same texture as regular slaw or pickles.

FREEZING FRUITS

To have a first class frozen product, you must select first class fruits to freeze. They will not improve while they are stored in the freezer.

Selecting

Fruits should be firm for good texture, yet ripe for good flavor. When picked for freezing, fruits should be ripe enough to eat. Most varieties are satisfactory for freezing if the fruit is firm-ripe and the color, flavor, and texture are characteristic of the fruit. Generally, if the variety is good to eat fresh, it is good to freeze.

Preparing

Discard moldy, soured, underripe, or overripe fruit. Remove twigs and leaves. Wash fruit thoroughly to remove dirt and chemical residue before peeling. To wash berries and other small fruits, place them in a large strainer or a small-mesh wire basket. Dip them up and down in several changes of cold water until they are free of dirt and sand. Shallow layers of fruit wash best. Careful handling is necessary to prevent bruising. Drain the fruit well.

Pretreating

Light-colored fruits, such as peaches and apples, have a tendency to turn dark when exposed to air. It's best to peel only 1 or 2 quarts at a time. To help prevent fruit from darkening, put each piece immediately after peeling or slicing into one of the five following solutions. Let it sit until ready to pack, at least 15 minutes.

1. To 1 gallon of water, add 2 tablespoons of salt and 2 tablespoons of clear, distilled vinegar. Don't let fruit stay in longer than 20 minutes because it will absorb vinegar and salt. **Rinse** fruit before adding sugar or syrup.

2. To 1 gallon water, add 1 tablespoon citric acid. Fruit doesn't absorb citric flavor and doesn't need to be rinsed before adding sugar or syrup.

3. To 1 gallon water, add 1 tablespoon ascorbic acid or 6 vitamin C tablets. Fruit doesn't need to be rinsed.

4. Buy ascorbic acid **mixture** and follow manufacturer's directions. Fruit doesn't need to be rinsed.

5. Add ascorbic acid or ascorbic acid mixture to the syrup in which fruit will be packed (see Syrup Pack, pp. 184-185).

Ascorbic acid is vitamin C and is generally more expensive than ascorbic acid mixture. The mixture usually consists of ascorbic acid, citric acid, sucrose (sugar), and some minor ingredients. Don't get the proportions for ascorbic acid and ascorbic acid mixture confused. You can buy ascorbic acid and citric acid at a drugstore; ascorbic acid mixture is available at drugstores, grocery stores, and other similar markets.

Packaging

Fruits may be packed dry, dry with sugar, in a sugar syrup, or with a sugar substitute. The method you use should depend on the kind of fruit, how you plan to use it, and your personal preference.

Using sugar or a sugar syrup helps fruits keep their flavor, color, texture, and aroma. In fact, cut fruits will discolor and become off-flavored very quickly after defrosting unless packed with sugar or syrup.

Dry Pack. The dry pack is good for small whole fruits, such as berries, that make a good quality product when frozen without sugar. Simply pack the fruit into a moisture-vapor-resistant container, fasten the lid or seal securely, and freeze.

Tray Pack. A tray pack is a form of dry pack that makes the fruit easier to remove from the container. You actually freeze the fruit before packing it into the freezer container.

Spread a single layer of prepared fruit on a baking sheet or shallow tray and place it in the freezer. Do not allow the pieces to touch each other. When the fruit is frozen, promptly pack it in moisture-vapor-resistant containers and quickly return it to the freezer before it can begin to thaw. The fruit pieces will remain loose rather than freezing together. You can remove the amount you need from the container without defrosting the whole package. Be sure to package the fruit as soon as it is frozen to prevent freezer burn.

Sugar Pack. Sugar draws juice from the fruit. Sprinkle it over the fruit until the sugar dissolves. Taste the juice and add more sugar, if desired. You can add ascorbic acid mixture to the sugar; proportions are listed on the container label. Or, the ascorbic acid mixture can be dissolved in ¼ cup water and mixed with the fruit before the sugar is added.

Pack fruit tightly in moisture-vapor-resistant freezer containers to prevent air pockets from forming. Don't add citric acid or lemon juice; they may mask the natural flavors or make the fruit too sour.

Syrup Pack. Fruits packed in a sugar syrup usually keep their normal size and shape better than those packed in sugar. A 40-percent syrup (Table 17, p. 185) is recommended for most fruits. You need ½ to ⅔ cup of syrup for each pint package of fruit.

Table 17. Syrups for Freezing

Syrup Type	Approx. % Sugar	Cups of Sugar	Cups of Water	Cups of Syrup Made
Very Light	10%	½	4	4¼
Light	20%	1	4	4½
Medium	30%	2	4	5
Medium Heavy	40%	3	4	5½
Heavy	50%	4¾	4	6½
Very Heavy	60%	7	4	7¾

To Make Syrup: Dissolve sugar in cold or hot water. If hot water is used, cool syrup before adding to fruit. Syrup may be prepared the day before and kept cold in the refrigerator. To help prevent fruit from darkening, add ½ teaspoon ascorbic acid to each quart of cold syrup just before pouring syrup over fruit. Follow directions on label when using ascorbic acid mixture. Don't add citric acid or lemon juice to syrup; they may mask natural flavors or make fruit too sour.

Note: Generally, up to ¼ of the sugar may be replaced by light corn syrup. For mild-flavored fruits, use the lighter syrups to prevent masking the fruit's natural flavor. Heavier syrups may be needed for very sour fruits.

When packing fruit into freezer containers, be sure the syrup completely covers fruit so that top pieces won't change in color and flavor. To keep fruit under syrup, place a small piece of clean, crumpled water-resistant paper, such as parchment, on top of fruit. Then, press fruit down into syrup before fastening lids or seals on containers.

Unsweetened Liquid Pack. Fruit can be packed in enough water, unsweetened fruit juice, or pectin syrup to cover the fruit. Use a small piece of clean, crumpled water-resistant paper, such as parchment, to hold fruit under the syrup.

Fruit frozen in unsweetened packs generally do not have the plump texture and good color of those packed with sugar. The fruits freeze harder and take longer to thaw. However, some fruits, such as raspberries, blueberries, gooseberries, steamed apples, rhubarb, and figs, make a good quality product when frozen without sugar.

Pectin Syrup. This is an unsweetened syrup often used for juicier fruits, such as strawberries or peaches. They retain their texture better in a syrup than if frozen in water or juice.

To make pectin syrup, combine 1 package of powdered pectin and 1 cup of water in a saucepan. Heat to boiling and boil 1 minute. Remove pan from heat and add 1¾ cups water. Stir and cool. This makes about 3 cups of moderately thick syrup. Add more water for a thinner syrup.

Sugar Substitutes. Non-caloric sweeteners can be added to any of the liquids used in unsweetened packs—water, juice, or pectin syrup. They may be used when freezing the fruits, or they may be added to the fruits just before serving. Sugar substitutes give a sweet flavor without added calories, but they do not furnish the beneficial effects of sugar, such as color protection and thickness of syrup.

Labels on the products give the equivalents to a standard amount of sugar. Use the manufacturer's directions on the container to determine the amount of sweetener to use.

Thawing

When thawed, fruits are ready to be served. Thaw only enough for one meal at a time. Fruits that have been frozen lose their freshness quickly after they have thawed. The texture softens and the flavor and color change.

Berries that have been frozen taste better when they still contain a few ice crystals. Though texture of peaches and similar fruits is better when they are still a little icy, flavor is improved by complete thawing.

For best flavor and color, always leave fruit in the **unopened** container during thawing. Turn the package several times during thawing to keep fruit coated with syrup and to help prevent darkening.

On the refrigerator shelf, it may take as long as 4 to 6 hours to thaw a 1-pound package of fruit. At room temperature, the time is shortened to 1 or 2 hours, depending on the pack. However, slow thawing in the refrigerator gives a better product. Fruits packed with dry sugar thaw more quickly than fruits packed in syrup.

You can refreeze thawed fruits if they have not reached 40°F and have not been partially thawed for longer than 1 to 2 days. There will be some loss of color, texture, and flavor when fruits are refrozen. You may prefer to make jam or preserves or to use thawed fruits in cooking rather than refreezing them.

Fruits usually ferment when they start to spoil. A little fermentation will not make fruits dangerous to eat, but it may spoil their flavor. Be sure any thawed fruit tastes and smells good before refreezing. **Whenever in doubt about thawed frozen foods, never refreeze and never eat.**

FRUIT RECIPES

APPLE SLICES

Use full-flavored apples that are crisp and firm, not mealy in texture. See General Steps in Freezing (pp. 169-170). A syrup pack is best for apples that will not be cooked before serving; sugar or dry pack is good for apples that will be used in pies. Wash apples.

Syrup Pack: Prepare a 40-percent syrup (see Table 17, p. 185). To keep apples from darkening, add ½ teaspoon ascorbic acid to each quart of syrup.

Pour ½ cup syrup into each pint container. Peel and core apples. Slice them into uniform sizes: about 12 slices for medium-size apples; 16 for large ones. Slice apples directly into the syrup in containers. Press apples down and add enough syrup to cover them. Leave headspace according to Table 14 (p. 179). Place crumpled paper on top to hold apples under syrup. Fasten lids securely and freeze.

Sugar or Dry Pack: Peel, core, and slice apples into uniform sizes. To prevent darkening, steam blanch (see Steam Blanching, p. 195) slices in a single layer 1½ to 2 minutes. Cool in cold water and drain. Or, sprinkle fruit with ½ teaspoon ascorbic acid mixed with 3 tablespoons water. Measure or weigh apple slices. Mix ½ cup sugar with each quart (1¼ pounds) of apples. Pack apples into containers, pressing fruit tightly. Leave a headspace according to Table 14 (p. 179). Or, omit sugar for dry pack. Fasten lids or seals securely and freeze.

Note: Apples are better canned than frozen.

BANANA PUREE

Use fully ripe bananas with peels that are flecked with brown. See General Steps in Freezing (pp. 169-170). Make a light syrup (Table 17, p. 185) with cold water or use hot water and allow time for it to cool. Peel bananas and mash or puree them, working quickly to prevent darkening. Add ascorbic acid powder to cool syrup, ½ teaspoon powder to each pint of syrup. Then, add syrup to pureed bananas, 1 or 2 tablespoons syrup for each banana used. Mix well.

Dry Pack: Spoon puree into clean plastic ice cube trays. Cover the trays tightly with cellophane freezer wrap, pliofilm, or polyethylene paper to prevent discoloration of the surface while puree is freezing. When frozen, remove puree cubes immediately and wrap each snugly with cellophane and seal. Pack the wrapped cubes tightly in freezer containers. Fasten lids or seals securely and return to the freezer.

BERRIES
(Except Blueberries and Strawberries)

Use fully ripe, firm berries. See General Steps in Freezing (pp. 169-170). Wash berries carefully in cold water and discard any that are soft, underripe, or defective. Drain well.

Syrup Pack: Prepare a 40- or 50-percent syrup (Table 17, p. 185) with cold water. Pack berries into containers and cover with syrup, leaving a headspace according to Table 14 (p. 179). Place crumpled paper on top to hold berries under syrup. Fasten lids securely and freeze.

Sugar or Dry Pack: Gently mix ¾ cup sugar with 1 quart (1⅓ pounds) berries. Pack berries in containers, leaving a headspace according to Table 14 (p. 179). Or, omit sugar for dry pack. Fasten lids or seals securely and freeze.

Tray Pack: Place berries in a single layer on a baking sheet or tray. Berries should not be crowded on tray. Place them in the freezer. When berries are frozen, pack them in containers. Fasten lids or seals securely and return to freezer.

BLUEBERRIES, HUCKLEBERRIES, OR ELDERBERRIES

Use full-flavored, ripe berries. See General Steps in Freezing (pp. 169-170). Remove leaves, stems, and immature or defective berries. Put fresh berries in a colander and quickly rinse with cold water. Pat the fruit dry with paper towels. A colander is preferred because water drains off the fruit quickly. Blueberries should never be allowed to stand in water. If they do, the skin may become tough.

Tray Pack: Place berries in a single layer on a baking sheet or tray. Berries should not be crowded on the tray. Place them in the freezer. When frozen, pack berries into containers. Fasten lids or seals securely and return to freezer.

Dry Pack: Pack berries into containers, leaving a headspace according to Table 14 (p. 179). Fasten lids or seals securely and freeze.

SOUR RED CHERRIES

Use bright red, tree-ripened cherries. See General Steps in Freezing (pp. 169-170). Wash cherries; remove stems and pits.

Syrup Pack: Pack cherries into containers and cover with a cold 50- to 60-percent syrup (Table 17, p. 185), depending on tartness of cherries. Leave a headspace according to Table 14 (p. 179). Place crumpled paper on top to hold cherries under syrup. Fasten lids securely and freeze.

Sugar Pack: Mix ¾ cup sugar with 1 quart (1⅓ pounds) cherries. Stir until sugar dissolves. Pack cherries into containers, leaving a headspace according to Table 14 (p. 179). Fasten lids or seals securely and freeze.

SWEET RED CHERRIES

Use bright, fully ripened cherries of dark-colored varieties. See General Steps in Freezing (pp. 169-170). Wash cherries and remove stems.

Syrup Pack: Prepare a 40-percent syrup (Table 17, p. 185), using cold water. Add ½ teaspoon ascorbic acid to each quart of syrup. Pour ½ cup cold syrup into each pint container, 1 cup in quarts. Remove pits and drop cherries immediately into syrup in containers. Press cherries down and add enough syrup to cover cherries. Leave a headspace according to Table 14 (p. 179). Place crumpled paper on top to hold cherries under syrup. Fasten lids securely and freeze.

FIGS

Use fully ripe figs. See General Steps in Freezing (pp. 169-170). Wash figs and remove stems.

Syrup Pack: Prepare a 40-percent syrup (Table 17, p. 185). Add ¾ teaspoon ascorbic acid or ½ cup lemon juice to each quart of syrup. Pour ½ cup syrup into each pint container, 1 cup in quarts. Peel figs, if desired, and cut them in halves or slices. Pack figs immediately as you peel and cut them into syrup in containers. Or, you can freeze them whole. Press figs down and add enough syrup to cover figs. Leave a headspace according to Table 14 (p. 179). Place crumpled paper on top to hold figs under syrup. Fasten lids securely and freeze.

Tray Pack: Peel figs, if desired, and cut them in halves or slices. Or, you can freeze figs whole. Measure prepared figs. Sprinkle each quart of figs with ¾ teaspoon ascorbic acid mixed in 3 tablespoons cold water. Place figs in a single layer on a baking sheet or tray. Do not crowd them on the tray. Place tray in the freezer. When frozen, pack figs in containers. Fasten lids or seals securely and return figs to the freezer.

Dry Pack: Prepare figs and sprinkle them with ascorbic acid in water same as for tray pack. Pack treated figs in containers, leaving a headspace according to Table 14 (p. 179). Fasten lids or seals securely and freeze.

GRAPEFRUIT SECTIONS

Use firm, tree-ripened grapefruit that feel heavy for their size and are free of soft spots. See General Steps in Freezing (pp. 169-170). Wash and peel grapefruit. Remove sections and discard all membranes and seeds. Save juice that accumulates while preparing sections.

Syrup Pack: Make a 40-percent syrup (Table 17, p. 185), using excess grapefruit juice for part of the water. Pack grapefruit sections into containers and cover with syrup, leaving a headspace according to Table 14 (p. 179). Place crumpled paper on top to hold fruit under syrup. Fasten lids securely and freeze.

For Juice: Squeeze juice from grapefruit, using a squeezer that does not press oil from the rind. Sweeten with 2 tablespoons sugar for each quart of juice. Or, pack without sugar. Pour juice into containers, leaving a headspace according to Table 14 (p. 179). If possible, pack juice in wide-mouth glass jars or citrus-enamel tin cans to avoid the development of an off-flavor. Fasten lids securely and freeze.

GRAPES

Use fully ripe, firm, sweet grapes. See General Steps in Freezing (pp. 169-170). Wash and remove stems. Discard any that are damaged or defective. Leave seedless grapes whole. Cut grapes with seeds in half and remove seeds.

Syrup Pack: Pack grapes into containers and cover with a cold 40-percent syrup (Table 17, p. 185). Leave a headspace according to Table 14 (p. 179).

Place crumpled paper on top to hold grapes under syrup. Fasten lids securely and freeze.

For Juice: Crush grapes and measure them into a large thick-bottom pot. Add 1 cup water for each gallon of crushed grapes. Heat to a simmer and simmer for 10 minutes. Strain juice through a jelly bag. To remove tartrate crystals, let juice sit overnight in the refrigerator. Carefully pour off the clear juice for freezing. Discard the sediment. Pour juice into containers, leaving a headspace according to Table 14 (p. 179). Fasten lids securely and freeze.

Note: If tartrate crystals form in frozen juice, they may be removed by straining the juice after it thaws.

MELON BALLS OR CUBES

Use ripe melons with firm flesh. See General Steps in Freezing (pp. 169-170). Wash melons, cut them in half, remove seeds, and peel. Cut melon flesh into slices, cubes, or balls.

Syrup Pack: Pack melon pieces into containers. Cover with a 30-percent syrup (Table 17, p. 185), leaving a headspace according to Table 14 (p. 179). Use the excess melon juice for some of the water, if available. Place crumpled paper on top to hold fruit under syrup. Fasten lids securely and freeze.

Note: Because of the large amount of water in melons, they make a poor quality frozen product. They are better if eaten partially frozen.

NECTARINE HALVES OR QUARTERS

Use well-ripened nectarines and handle carefully to avoid bruising. See General Steps in Freezing (pp. 169-170). Wash nectarines.

Syrup Pack: Prepare a 40-percent syrup (Table 17, p. 185). To keep nectarines from darkening, add ½ teaspoon ascorbic acid to each quart of syrup. Pour ½ cup syrup into each pint container; 1 cup in each quart. Peel nectarines and cut them into halves or quarters. Drop pieces immediately into syrup in containers. Press pieces down and add enough syrup to cover fruit, leaving a headspace according to Table 14 (p. 179). Place crumpled paper on top to hold nectarines under syrup. Fasten lids securely and freeze.

Sugar Pack: Peel nectarines and cut them into halves or quarters. Measure or weigh pieces. Add ⅔ cup sugar to each quart (1⅓ pounds) of prepared nectarines. Mix until sugar dissolves or let it sit for 15 minutes. Sprinkle each quart of nectarines with ¼ teaspoon ascorbic acid in 3 tablespoons cold water. Pack treated nectarines into containers, leaving a headspace according to Table 14 (p. 179). Fasten lids or seals securely and freeze.

Note: Nectarines are better canned than frozen.

ORANGE SECTIONS

Use firm, tree-ripened oranges that feel heavy for their size and are free of soft spots. See General Steps in Freezing (pp. 169-170). Wash and peel oranges. Remove sections and discard all membranes and seeds. Save juice that accumulates while preparing sections.

Syrup Pack: Make a 40-percent syrup (Table 17, p. 185), using excess orange juice for part of the water. Pack orange sections into containers and cover with syrup, leaving a headspace according to Table 14 (p. 179). Place crumpled paper on top to hold fruit under syrup. Fasten lids securely and freeze.

For Juice: Squeeze juice from oranges, using a squeezer that does not press oil from the rind. Sweeten with 2 tablespoons sugar for each quart of juice. Or, pack without sugar. Pour juice into containers, leaving a headspace according to Table 14 (p. 179). If possible, pack juice in wide-mouth glass jars or citrus-enamel tin cans to avoid the development of an off-flavor. Fasten lids securely and freeze.

PEACH HALVES OR QUARTERS

Use well-ripened peaches and handle them carefully to avoid bruising. See General Steps in Freezing (pp. 169-170). Wash peaches.

Syrup Pack: Prepare a 40-percent syrup (Table 17, p. 185). To keep peaches from darkening, add ½ teaspoon ascorbic acid to each quart of syrup. Pour ½ cup syrup into each pint container; 1 cup in each quart. Peel peaches and cut them into halves or quarters. Drop pieces immediately into syrup in containers. Press pieces down and add enough syrup to cover fruit, leaving a headspace according to Table 14 (p. 179). Place crumpled paper on top to hold fruit under syrup. Fasten lids securely and freeze.

Sugar Pack: Peel peaches and cut them into halves or quarters. Measure or weigh pieces. Add ⅔ cup sugar to each quart (1⅓ pounds) of prepared peaches. Mix until sugar dissolves or let it sit for 15 minutes. Sprinkle each quart of peaches with ¼ teaspoon ascorbic acid in 3 tablespoons cold water. Pack treated peaches into containers, leaving a headspace according to Table 14 (p. 179). Fasten lids or seals securely and freeze.

Note: Peaches are better canned than frozen.

PEAR SLICES

Use full-flavored, well-ripened pears that are crisp and firm, not mealy in texture. See General Steps in Freezing (pp. 169-170). Wash pears.

Syrup Pack: Prepare a 40-percent syrup (Table 17, p. 185). Peel, core, and slice pears into uniform sizes: 12 slices for medium-size pears; 16 for large ones. Drop slices immediately into boiling syrup and heat for 1 to 2 minutes, depending on size of pears. Drain pears and save syrup. Cool both fruit and syrup. Add ¾ teaspoon ascorbic acid to each quart of cold syrup. Pack pears in containers and cover with cold syrup, leaving a headspace according to Table 14 (p. 179). Place crumpled paper on top to hold pears under syrup. Fasten lids securely and freeze.

JAPANESE PERSIMMONS

See General Steps in Freezing (pp. 169-170). Wash persimmons and dry well. Remove stems.

Tray Pack: Place whole persimmons in a single layer on a baking sheet or tray; place them in the freezer. When frozen solid, wrap each persimmon individually in freezer foil or bags. Fasten seals securely and return to freezer.

CULTIVATED OR NATIVE PERSIMMONS

Use deep orange or red, soft-ripe persimmons. See General Steps in Freezing (pp. 169-170). Wash and peel persimmons. Cut them into sections and remove seeds.

For Puree: Press fruit through a sieve or puree in blender. Add ⅛ teaspoon ascorbic acid or 1½ teaspoons crystalline citric acid to each quart of puree. If using a cultivated variety, you may want to add 1 cup sugar to each quart (2 pounds) of puree. Pack puree into containers, leaving a headspace according to Table 14 (p. 179). Fasten lids or seals securely and freeze.

PLUMS

Use firm, ripe plums that are soft enough to yield to slight pressure. See General Steps in Freezing (pp. 169-170). Wash plums.

Syrup Pack: Prepare a 40- to 50-percent syrup (Table 17, p. 185), depending on tartness of plums. Add ½ teaspoon ascorbic acid to each quart of syrup. Pour ½ cup syrup into each pint container, 1 cup into quarts. Cut plums in half and remove pits. Drop them immediately into syrup in containers. Press plums down and add more syrup to cover them, leaving a headspace according to Table 14 (p. 179). Place crumpled paper on top to hold plums under syrup. Fasten lids securely and freeze.

Note: You can freeze plums whole in a syrup pack.

PLUM SAUCE

Use firm, ripe plums that are soft enough to yield to slight pressure. See General Steps in Freezing (pp. 169-170). Wash plums. Boil fully ripe clingstone plums without water until soft; then, remove pits and skins. Continue cooking pulp and juice until it thickens. Add 1 cup sugar to each quart of plums. Cool and fill containers, leaving a headspace according to Table 14 (p. 179). Fasten lids securely and freeze.

RHUBARB

Use firm, tender, well-colored stalks with good flavor and few fibers. See General Steps in Freezing (pp. 169-170). Wash rhubarb, trim and cut it into lengths that fit the container you will use. To prevent loss of color and flavor, blanch rhubarb in boiling water for 1 minute and cool promptly in a pan of cold water that contains ice cubes.

Syrup Pack: Prepare a 40-percent syrup (Table 17, p. 185) with cold water. Pack raw or blanched rhubarb tightly into containers. Cover rhubarb with syrup, leaving a headspace according to Table 14 (p. 179). Fasten lids securely and freeze.

Dry Pack: Pack raw or blanched rhubarb tightly into containers, leaving a headspace according to Table 14 (p. 179). Fasten lids or seals securely and freeze.

STRAWBERRIES

Use fully ripe, firm berries with a deep red color. Discard any that are immature or defective. See General Steps in Freezing (pp. 169-170). Wash strawberries carefully and drain well. Remove caps and stems. Leave berries whole, crush, or slice.

Syrup Pack: Prepare a 50-percent syrup (Table 17, p. 185) with cold water. Pack berries into containers and cover with cold syrup, leaving a headspace according to Table 14 (p. 179). Place crumpled paper on top to hold fruit under syrup. Fasten lids securely and freeze.

Sugar Pack: Measure or weigh prepared strawberries. Add ¾ cup sugar to each quart (1⅓ pounds) of berries. Mix thoroughly until sugar dissolves or let it sit for 15 minutes. Pack containers, leaving a headspace according to Table 14 (p. 179). Fasten lids or seals securely and freeze.

TOMATOES

See page 203.

FREEZING VEGETABLES

Most kinds of vegetables can be successfully frozen. When properly selected, prepared, frozen, and stored, they hold their fresh qualities. The flavor, color, texture, and nutritive value do not change. Vegetables which have been bought, taken home, and then frozen usually are not as fresh tasting as those which have been grown at home. Practically all frozen vegetables may be held in the freezer for about 1 year.

Foods properly frozen and stored come from the freezer just as good as when they were put in—but no better. Freezing does not improve the product.

Selecting

Use only young and tender vegetables that are at the right stage for eating fresh. Select the varieties that you enjoy eating fresh and that are especially suitable for freezing. Do not put too much importance on varieties; other factors are equally important.

Preparing

If possible, gather vegetables in the cool part of the morning. Prepare and freeze them with the least possible delay; vegetables deteriorate rapidly at room temperature. If it is impossible to process them promptly, place them in the refrigerator at a temperature of from 33° to 40°F for not more than a few hours. The shorter the time from harvesting to freezing, the better the product will be. Vegetables should not be prepared the night before and held until the next day before being frozen.

Wash all vegetables thoroughly to remove dirt and trash before removing skins or shells. Wash small amounts at a time. Use a large quantity of clean, cool water and change it frequently. It is often necessary to use a brush to remove dirt from potatoes, carrots, and other root vegetables. Float and stir greens so that leaves will not stick together. Washing also removes spray residues and lowers the number of spoilage bacteria in the finished product. While washing, remove any bruised, spoiled, or overmature vegetables that were overlooked when they were sorted. Do not let vegetables soak in water while washing.

Blanching

It is necessary to blanch almost all vegetables before freezing them. Only green peppers and onions that have been cut should not be blanched; whole onions do need blanching. Blanching slows or stops the action of enzymes that cause the loss of color, flavor, and texture in vegetables. Blanching helps retain vitamins. It also helps clean dirt and organisms from the surface, and it wilts or softens vegetables, making them easier to pack into containers.

Although vegetables that have not been blanched before freezing are usually safe to eat, their overall eating quality may be poor. Unblanched vegetables should be used within 2 months after freezing for acceptable quality.

Blanching time is crucial, and it varies with the vegetable and its size. See Table 25 (p. 248) or specific recipes for blanching times. Underblanching stimulates the activity of enzymes and is worse than no blanching. Overblanching causes loss of flavor, color, vitamins, and minerals.

Water Blanching. This method is recommended for most vegetables. It is done in large quantities of boiling water. You generally use 1 gallon of water for each pint (pound) of vegetables; use 2 gallons for each pound of leafy vegetable. Put the vegetable in a blanching basket and lower it into actively boiling water. Cover the blancher with a lid and start counting blanching time from the moment the water returns to boiling after adding the vegetable. Keep range unit on its highest temperature throughout blanching time.

If you do not have a blancher, you can use a large pot with a wire basket or a colander that will fit inside to hold the vegetables. The basket should be large enough for water to completely surround each individual piece of vegetable.

Promptly cool blanched vegetables (see below).

Steam Blanching. Steaming can be done in a blancher, a vegetable steamer, or a large pot with a tight-fitting lid. This method is recommended for broccoli, pumpkin, sweet potatoes, and winter squash. They each require 5 minutes of steam blanching. Place the vegetables—one layer deep—in a wire basket or on a rack 3 inches above 1 to 2 inches of rapidly boiling water. Cover the steamer or pot with the lid as quickly as possible. Begin timing when the steam rises again after vegetables are placed in the steamer. Promptly cool blanched vegetables (see below).

Steam blanching is tricky and less reliable than water blanching. Care must be taken to see that layers of vegetables are thin enough so that steam will reach all parts at almost the same time. This method is not satisfactory for leafy green vegetables because the leaves tend to mat together.

Cooling

As soon as blanching is complete, vegetables must be cooled quickly and thoroughly to stop the cooking action. Cool them by holding the blancher basket under cold running water or by plunging it immediately into a pan of cold water with ice cubes in it. Use 1 pound of ice for each pound of vegetable. A water temperature of 60°F or lower is necessary for cooling. Cooling blanched vegetables should take about the same amount of time as blanching them.

When the vegetable is cool, remove it from the water and drain thoroughly. Don't let the vegetable sit in water longer than necessary because it will lose flavor and nutrients.

To cool cooked vegetables such as pumpkin, tomatoes, etc., pour them from the hot pan they were cooked in into a cool pan. Then, put the cool pan into a larger one filled with ice and ice water.

Packaging

Immediately after the vegetables have been blanched, cooled, and drained, pack them in meal-size, moisture-vapor-resistant freezer bags or rigid containers. Vegetables can be packed in one of two ways for freezing.

Dry Pack. Pack the vegetable tightly in the containers and remove as much air as possible. Leave the headspace specified in the recipe and fasten the lids or seals securely. Label the package.

Tray Pack. Spread the vegetable in a single layer on a baking sheet or shallow tray. Place it in the freezer just long enough for the vegetable to freeze firm. When frozen, pack the vegetable tightly in the container and remove as much air as possible. Leave the headspace specified in the recipe and fasten the lids or seals securely. Label the package.

Thawing and Cooking

Blanching vegetables before freezing partially cooks them. And, the frost on frozen vegetables furnishes some moisture. Therefore, frozen vegetables will cook to the recommended tender-crisp stage in less time and in less water than fresh vegetables.

You can cook most frozen vegetables without thawing; but corn on the cob must be completely thawed. Some vegetables, such as greens, asparagus, and broccoli, cook more evenly if thawed just enough to separate the leaves or stalks. Never refreeze vegetables that have completely thawed.

Thaw and cook only enough frozen vegetable for one meal at a time. If you cook more than one package at a time or one large package, use a wide pan to reduce the cooking time. Use only ¼ to ½ cup water for four to five servings, depending on the size of the package, the time needed to cook the vegetable, and the type vegetable.

Bring lightly salted water to a boil. Add frozen vegetable, cover, and bring back to a rapid boil. Use a tight-fitting lid so that the moisture will form steam for cooking. Then reduce heat, but keep it high enough for the water to boil steadily. Cook vegetables only until tender; overcooking destroys flavor, texture, and nutritive value.

Vegetables may also be microwaved. Follow the recommendations in the user's manual that came with your oven.

VEGETABLE RECIPES

ASPARAGUS SPEARS

Use young tender spears. See General Steps in Freezing (pp. 169-170). Wash asparagus well and sort spears by thickness. Trim the stalks by removing the scales with a sharp knife. Cut them into lengths that will fit the freezer container. Blanch small spears 2 minutes in boiling water; medium spears, 3 minutes; and large spears, 4 minutes. Cool and drain. Pack asparagus into containers, leaving no headspace. Fasten lids or seals securely and freeze.

LIMA BEANS
(Butter, Pinto, or Soy)

Use beans picked while the seeds are in the green stage. See General Steps in Freezing (pp. 169-170). Wash beans well; shell and sort them according to size. Blanch small beans 2 minutes in boiling water; medium beans, 3 minutes; and large beans, 4 minutes. Cool and drain. Then, sort out any beans that turned white, if desired. Pack beans into containers, leaving a ½-inch headspace. Fasten lids or seals securely and freeze.

SNAP BEANS
(Green, Italian, or Wax)

Use young tender pods picked when the seeds are first formed. See General Steps in Freezing (pp. 169-170). Wash beans well in cold water. Trim off stems and tips. Leave beans whole, slice lengthwise, or cut into 2-inch pieces. Blanch 3 minutes in boiling water. Cool and drain. Pack beans into containers, leaving a ½-inch headspace. Fasten lids or seals securely and freeze.

BEET SLICES OR CUBES

Use deep red, tender, young beets. See General Steps in Freezing (pp. 169-170). Wash beets well. Trim tops but leave ½-inch stems and roots to prevent bleeding of color during cooking. Drop beets into a large pan of boiling water and boil small beets 25 to 30 minutes; larger beets, 45 to 50 minutes. Cool and drain. Peel beets, remove stem and root, and slice or cut into cubes, if desired. Pack beets into containers, leaving a ½-inch headspace. Fasten lids or seals securely and freeze.

BROCCOLI

Use firm, young, tender stalks with compact heads. See General Steps in Freezing (pp. 169-170). Cut off large leaves and tough stalks. Separate heads into convenient-size sections. To remove insects, soak broccoli with heads down in salt water (4 teaspoons salt to 1 gallon water) for about 30 minutes. Discard soaking water. Split lengthwise so heads are not more than 1½ inches across. Blanch 5 minutes in steam or 3 minutes in boiling water. Cool and drain. Pack broccoli into containers, leaving no headspace. Fasten lids or seals securely and freeze.

CABBAGE OR CHINESE CABBAGE

Use freshly picked, solid heads. See General Steps in Freezing (pp. 169-170). Trim off coarse outer leaves. Cut heads into medium shreds or thin wedges. Blanch 1½ minutes in boiling water; cool and drain. Pack into containers, leaving a ½-inch headspace. Fasten lids or seals securely and freeze.

Note: Use frozen cabbage only as a cooked vegetable.

CARROTS

Use young, tender carrots that do not have cores. See General Steps in Freezing (pp. 169-170). Remove tops, wash well, and scrape or thinly peel. Cut into ¼-inch slices or lengthwise strips. Leave small carrots whole. Blanch slices and strips 2 minutes in boiling water. Blanch small, whole carrots 5 minutes. Cool and drain. Pack carrots into containers, leaving a ½-inch headspace. Fasten lids or seals securely and freeze.

CAULIFLOWER

Use white, compact heads. See General Steps in Freezing (pp. 169-170). Break or cut flowerlets into pieces about 1 inch across. Wash well. To remove insects, soak cauliflower 30 minutes in salt water (4 teaspoons salt to 1 gallon water). Discard soaking water. Blanch cauliflower 3 minutes in boiling salt water (4 teaspoons salt to 1 gallon water). Cool and drain. Pack cauliflower into containers, leaving no headspace. Fasten lids or seals securely and freeze.

CREAM-STYLE CORN

Use tender, freshly picked corn in the milk stage. See General Steps in Freezing (pp. 169-170). Remove husks and silks and trim off bad spots. Wash well and sort by size. Cut off tips of kernels and scrape the corn from the cob. Add ½ to 1 pint of water to each pint of corn and boil 4 minutes. Cool by scraping corn into a cold pan and placing pan in a larger pan of ice water. Pack corn into containers, leaving a ½-inch headspace. Fasten lids or seals securely and freeze.

CORN ON THE COB

Use tender, freshly gathered corn in the milk stage. See General Steps in Freezing (pp. 169-170). Remove husks and silks and trim off bad spots. Wash well. Blanch small ears 7 minutes in boiling water; medium ears, 9 minutes; and large ears, 11 minutes. Cool completely to prevent corn from tasting like the cob. Drain well. Pack corn into containers, leaving no headspace. Fasten lids or seals securely and freeze.

WHOLE-KERNEL CORN

Use tender, freshly gathered corn in the milk stage. See General Steps in Freezing (pp. 169-170). Remove husks and silks and trim off bad spots. Wash well. Blanch corn on the cob 4 minutes in boiling water. Cool and drain. Cut kernels off cob about two-thirds the depth of the kernels. Pack kernels into containers, leaving a ½-inch headspace. Fasten lids or seals securely and freeze.

EGGPLANT SLICES

Use eggplant that was harvested before seeds became mature and color is uniformly dark. See General Steps in Freezing (pp. 169-170). Wash eggplant well. Peel and cut it into ⅝-inch slices. Prepare eggplant quickly to prevent darkening; prepare only enough for one blanching at a time. To 1 gallon boiling water, add 4½ teaspoons citric acid or ½ cup lemon juice. Blanch 4 minutes in boiling water. Cool and drain. Pack eggplant into containers, leaving a ½-inch headspace. Fasten lids or seals securely and freeze.

Note: If slices will be fried, pack with two pieces of freezer paper between slices to make them easier to separate.

GREENS
(Collards, Mustard, Spinach, or Turnips)

Use young, tender green leaves. See General Steps in Freezing (pp. 169-170). Wash greens thoroughly. Remove imperfect leaves and large, tough stems. Blanch collards 3 minutes in boiling water; blanch all other greens 2 minutes. Use 2 gallons of water for each pint of greens. Cool and drain. Pack greens into containers, leaving a ½-inch headspace. Fasten lids or seals securely and freeze.

FRESH HERBS

Use fresh, tender leaves and stems. See General Steps in Freezing (pp. 169-170). Wash herbs thoroughly in cool water to remove any grit. Drain and pat dry with paper towels. Wrap small amounts of sprigs or leaves in freezer wrap and place them in a freezer bag. This will allow you to take out only the amount needed for a recipe. Fasten seals securely and freeze.

Note: Frozen herbs usually are not suitable as a garnish because they become limp when thawed. They may be chopped and used in cooking.

OKRA

Use young tender pods. The smooth varieties do not split as easily as the ridged varieties. See General Steps in Freezing (pp. 169-170). Wash okra well and separate pods that are 4 inches long or under from larger pods. Remove the stems at the end of the seed cells, being careful not to expose the cells. Blanch the smaller pods 3 minutes in boiling water; larger pods, 4 minutes. Cool and drain. Pack okra into containers, leaving a ½-inch headspace. Fasten lids or seals securely and freeze.

MEALED OKRA SLICES

Use young tender pods. The smooth varieties do not split as easily as the ridged varieties. See General Steps in Freezing (pp. 169-170). Wash okra well and separate pods that are 4 inches long or under from larger pods. Remove the stems at the end of the seed cells, being careful not to expose the cells. Blanch the smaller pods 3 minutes in boiling water; larger pods, 4 minutes. Cool and drain. Slice blanched okra crosswise and coat with corn meal or flour. Spread slices in a single layer on a baking sheet or tray. Place it in the freezer just long enough for okra slices to freeze firm. Pack okra in containers quickly, leaving a ½-inch headspace. Fasten lids or seals securely and return to freezer.

CHOPPED ONIONS

Use mature onions. See General Steps in Freezing (pp. 169-170). Clean onions as for eating and cut into ¼-inch pieces. Pack onions into containers, leaving no headspace. Fasten lids or seals securely and freeze.

Note: Use frozen onions within 1 month.

GARDEN PEAS
(English or Green)

Use peas that were picked when pods were filled with young, tender peas that had not become starchy. See General Steps in Freezing (pp. 169-170). Wash pods to remove loose dirt and shell peas. Blanch them 1½ minutes in boiling water. Cool and drain well. Pack peas into containers, leaving a ½-inch headspace. Fasten lids or seals securely and freeze.

SOUTHERN PEAS
(Blackeye, Crowder, or Field)

Use well-filled pods that were picked when peas were tender. See General Steps in Freezing (pp. 169-170). Wash pods well. Shell peas and discard any immature or tough ones. Blanch peas 2 minutes in boiling water. Cool and drain. Pack peas into containers, leaving a ½-inch headspace. Fasten lids or seals securely and freeze.

SUGAR SNAP PEAS

Use unblemished pods in which the seeds have not yet developed. See General Steps in Freezing (pp. 169-170). Wash pods well, remove strings, and trim ends. Leave pods whole or cut them into 1-inch pieces. Boil enough water in a large saucepan to cover pods. Also, prepare a large pan of water and ice. Add pods to boiling water and blanch 1 minute for small pods; 1½ to 2 minutes for large ones. Drain immediately and cool pods in ice water for the same amount of time. Drain again and pack pods into containers, leaving a ½-inch headspace. Fasten lids or seals securely and freeze.

To Vary: Pods may be tray packed after being blanched and cooled. Place pods in a single layer on a baking sheet or tray. Do not let them overlap. Place them in the freezer. As soon as pods have frozen, pack them in containers. Fasten lids or seals securely and return to freezer.

HOT PEPPERS
(Banana, Chili, Hungarian, or Jalapeno)

Use tender, well-colored peppers. See General Steps in Freezing (pp. 169-170). Wash peppers and remove stems. Pack peppers into containers, leaving no headspace. Fasten lids or seals securely and freeze.

Caution: Wear rubber gloves while handling chilies and wash hands thoroughly with soap and water before touching your face.

SWEET PEPPERS

Use firm, tender, crisp, thick-walled peppers. See General Steps in Freezing (pp. 169-170). Wash peppers and cut them in half. Cut out stems, membrane, and seeds. Cut them into strips or rings or chop them into ¼- to ½-inch pieces, if desired. Or, freeze them as halves or quarters. Pack peppers into containers, leaving no headspace. Fasten lids or seals securely and freeze.

Note: You can water blanch strips or rings for 2 minutes and halves or quarters for 3 minutes. Cool and drain before packing, leaving a ½-inch headspace. Blanched peppers are good for use in cooking.

PIMIENTOS

Use firm, crisp pimientos with deep red color. See General Steps in Freezing (pp. 169-170). Peel pimientos by roasting them on a baking sheet at 400°F until skins crack and blister (5 to 10 minutes). Or, boil them 3 to 5 minutes until skins slip easily. Wash off the charred skins, cut out the stems, and remove seeds. Rinse, cool, and drain pimientos. Pack them into containers, leaving a ½-inch headspace. Fasten lids or seals securely and freeze.

NEW IRISH POTATOES

Use smooth, freshly picked new potatoes. See General Steps in Freezing (pp. 169-170). Wash potatoes well and peel or scrape them. Blanch small potatoes (¾ inch in diameter) 4 minutes in boiling water; 1-inch potatoes, 6 minutes; and 1½-inch potatoes, 7 minutes. Cool and drain. Pack potatoes into containers, leaving a ½-inch headspace. Fasten lids or seals securely and freeze.

MASHED PUMPKIN

Use full-colored, mature pumpkin with fine texture. See General Steps in Freezing (pp. 169-170). Wash the pumpkin well and cut it into cooking-size pieces. Remove seeds. Cook pumpkin in a small amount of boiling water until it is tender. Remove pulp from rind and mash. Put hot mashed pumpkin in a cool pan and place it in ice water to cool. Pack pumpkin into containers, leaving a ½-inch headspace. Fasten lids or seals securely and freeze.

SPAGHETTI SQUASH

Use fully mature squash. See General Steps in Freezing (pp. 169-170). Wash squash and cut it in half crosswise. Scoop out seeds. Boil 2 to 3 inches of water in a wide pan. Place squash halves on a rack over boiling water with the cut side down. Cover pan and steam about 20 minutes or until tender. Remove halves and pull out spaghetti-like strands. Drain and cool thoroughly. Pack spaghetti strands into containers. Fasten lids or seals securely and freeze.

SUMMER SQUASH SLICES

Use young, tender squash. See General Steps in Freezing (pp. 169-170). Wash squash well and cut it into ½-inch slices. Blanch squash 3 minutes in boiling water. Cool and drain. Pack squash into containers, leaving a ½-inch headspace. Fasten lids or seals securely and freeze.

SWEET POTATO SLICES

Use medium to large potatoes that have been cured at least one week. See General Steps in Freezing (pp. 169-170). Wash potatoes and sort according to size. Peel and place potatoes immediately in salt water (½ cup salt to 1 gallon water) to prevent darkening. Cut treated potatoes into ½-inch slices. Blanch slices 5 minutes in boiling water. Cool and drain. Pack potatoes into containers, leaving a ½-inch headspace. Cover slices with syrup (equal parts sugar and water), leaving a ½-inch headspace. Place crumpled water-resistant paper, such as parchment, on top to hold potatoes under syrup. Fasten lids or seals securely and freeze.

MASHED SWEET POTATOES

Use medium to large sweet potatoes that have been cured for at least one week. See General Steps in Freezing (pp. 169-170). Wash potatoes well. Boil or bake them in their skins until tender. Let them cool at room temperature; then, peel and mash them. To prevent darkening, add 2 tablespoons orange or lemon juice to each quart of mashed potatoes. Sugar may be added before freezing. Pack potatoes into containers, leaving a ½-inch headspace. Fasten lids or seals securely and freeze.

TOMATOES

Use firm, ripe tomatoes with deep red color. See General Steps in Freezing (pp. 169-170). Remove stem ends; peel and quarter tomatoes. Place quarters in a saucepan, cover, and cook until tender, 10 to 20 minutes. Pour tomatoes into a cool pan and place it in a larger pan of ice and cold water to cool tomatoes. Pack tomatoes into containers, leaving a ½-inch headspace in pints or 1-inch in quarts with wide openings. If using glass jars with a narrow-top opening, leave a ¾-inch headspace in pints or 1½-inch in quarts. Fasten lids or seals securely and freeze.

GREEN TOMATO SLICES

Use firm, sound green tomatoes. See General Steps in Freezing (pp. 169-170). Wash, core, and cut tomatoes into ¼-inch slices. Pack slices into containers with two pieces of freezer paper between slices. Leave a ½-inch headspace. Fasten lids or seals securely and freeze.

FREEZING MEAT, POULTRY, AND SEAFOOD

Freezing doesn't improve the flavor, texture, or color of any animal product. If poor quality meat, poultry, or fish is put in the freezer, a poor quality product will be taken out. If a good quality product is put into the freezer but is not prepared, packaged, frozen, and stored correctly, a poor quality will be the result. Remember, one step is just as important as another.

Meat
(Beef, Lamb, Veal, or Pork)

Read General Steps in Freezing (pp. 169-170) before preparing meat to freeze.

Selecting. Start with a good, healthy animal that is about the right size for the cuts you want (see Table 7, p. 14). Do not feed the animal for 24 hours before slaughter, but give it adequate water. Exciting or overheating an animal before it is slaughtered may prevent it from bleeding freely. Sanitary conditions are essential to reduce the growth of food spoilage bacteria.

Slaughtering. Properly prepared meat keeps better. Whether slaughtering is done at home or at a locker plant, a good clean job of dressing is important. Thorough bleeding is very necessary. It is generally best to have the animal killed, dressed, aged, packaged, and frozen at a reliable freezer-locker plant.

Chilling. Prompt chilling of a warm, dressed carcass is essential. Carcasses should be hung in a chill room so that they do not touch each other. This allows free air circulation for removal of heat. They should be cooled to a temperature of below 40°F within 24 hours to prevent souring or spoiling.

Aging. This is a process of holding meat at a temperature low enough to delay growth of spoilage bacteria, but still high enough to permit the natural enzymes in the meat to soften the connective tissues, making it more tender and flavorful. To age beef and lamb, hold it for 5 to 7 days after slaughter at 34°F in a chill room or, preferably, in a special aging room. Age veal and pork for 1 to 2 days.

Cutting. After meat has been properly chilled and aged, cut it into pieces best suited to your needs. See General Steps in Freezing (pp. 169-170). Meat may be cut into roasts, rolled roasts, steaks, chops, stew meat, or ground meat. Compact, smoothly trimmed cuts are the easiest to wrap.

Wrapping. Package meats in meal-size portions. Remove as many bones as possible because they take up freezer space. Use only special moisture-vapor-resistant wrapping material or freezer containers (see Packaging Materials, pp. 171-172, and Packaging Food, pp. 175-179). The drugstore wrap (p. 177) and the butcher wrap (p. 178) make airtight packages for meats. Two sheets of wax paper between cuts such as steaks and chops will allow you to separate them without thawing. Ground meat and stew meat are often packaged in moisture-vapor-resistant bags with cover boxes or rigid plastic freezer containers. However you package your meat, be sure to press all air out of the package and seal it properly. An overwrap of butcher paper and an inner wrap of freezer material will prevent punctures.

Freezing. As soon as meat has been properly wrapped, place it in the coldest part of the freezer and freeze at 0°F or lower. Spread packages so that air can circulate between them. Most cuts will be frozen in 12 to 24 hours at 0°F or below. Do not put in the freezer, at any one time, more than 2 to 3 pounds of unfrozen meat per cubic foot of freezer space. **If you have a 20-cubic-foot home freezer, never add more than 40 to 60 pounds of unfrozen meat at one time.** When this amount is completely frozen, more unfrozen food can be added. Always place unfrozen food in coldest part of the freezer and leave air space around each package until frozen. Check the user's manual that came with your freezer to know where the coldest places are.

Thawing and Cooking. The refrigerator is the best place to thaw frozen meat. The meat is less likely to spoil, develop an off-flavor, or seep juices. Leave the meat wrapped. Be sure to allow enough time for thawing. It generally takes about 4 hours for each pound; however, this varies according to meat thickness, amount of bone, and temperature of the refrigerator. Meat allowed to thaw at normal room temperature is more susceptible to food spoilage.

If frozen meat is first thawed, the cooking time is about the same as for fresh meat. Meat which has not thawed takes a longer cooking period than fresh meat.

All meats, fresh or frozen, should be cooked at moderately low temperatures—350°F or lower. This is especially important when cooking meat that is frozen solid. Otherwise, the meat will not cook evenly throughout.

Wild Game

Deer, moose, antelope, and other large game should be prepared and packaged for freezing the same as beef, lamb, veal, and pork, above. See General Steps in Freezing (pp. 169-170). Trim and discard bloodshot meat before freezing.

Rabbit, squirrel, and other small game should be skinned, dressed, and then chilled. Refrigerate them for 24 to 36 hours until the meat is no longer rigid. Then, cut into serving-size pieces or leave whole. Package, seal, and freeze.

Poultry

Read General Steps in Freezing (pp. 169-170) before preparing poultry to freeze.

Selecting. Use only fresh, high quality poultry to freeze. The tender young bird is best for roasting, frying, and broiling. The more flavorful older birds are better braised or stewed.

Packaging. Package poultry in freezer paper, using the drugstore wrap (p. 177) or butcher wrap (p. 178). Or, you can package poultry in freezer bags. When packaging pieces, arrange them to form a compact, square, flat package so they will stack better in the freezer. After packaging, seal and freeze immediately.

Store-bought poultry needs to be overwrapped before freezing because the clear wrap used by grocery stores is not moisture-vapor-resistant.

Note: Do not stuff poultry, turkey, or game birds before freezing them. During freezing or thawing, food poisoning bacteria can easily grow in the stuffing. Commercially stuffed frozen poultry is prepared under special safety conditions that cannot be duplicated at home.

Thawing. Thaw poultry products in the refrigerator, not at room temperature. Or, they may be thawed under continuously running water.

Game Birds

Quail, dove, duck, pheasant, and other game birds should be gutted and dressed as soon as possible after shooting. Cool and clean them thoroughly. Remove excess fat from wild ducks and geese because it becomes rancid very quickly. Freeze the same as for poultry, (pp. 205-206).

Seafood

Read General Steps in Freezing (pp. 169-170) before preparing seafood to freeze.

Preparing. Gut and thoroughly clean fish soon after they are caught. Prepare fish the same way you will use them. Cut large fish into steaks or fillets. Steaks are cut across the backbone at 1-inch intervals. Fillets are boneless portions cut lengthwise away from the backbone. Freeze small fish whole after they are cleaned.

Packaging. Package fish in meal-size servings in moisture-vapor-resistant aluminum foil, freezer wrap, or freezer bags. If several pieces of fish are placed in the same package, separate them with two pieces of wax paper for easier thawing. Store the fish at 0°F or lower. Glazing the fish before wrapping and freezing will decrease drip loss when it thaws. It also firms the fish. Fish may be glazed using any of the following methods.

Lemon Gelatin Glaze. Mix ¼ cup lemon juice with 1¾ cups water. Pour ½ cup of the mixture in a small bowl and sprinkle one packet of unflavored gelatin over the top. Heat the remaining 1½ cups of mixture to boiling. Stir the gelatin mixture into the boiling mixture until gelatin completely dissolves. Cool the mixture to room temperature. When cool, dip fish into the lemon-gelatin glaze and drain. Package, label, and freeze fish.

Ice Glaze. Place unwrapped fish on a baking sheet in the freezer. Prepare a pan of ice water and chill it to near-freezing. As soon as fish are frozen, dip them in the ice water. Place the fish on the pan in the freezer again for a few minutes to harden the glaze. Dip the fish in the ice water again and repeat the process until a uniform cover of ice is formed over each fish. Package, label, and store fish in freezer.

Water Pack. Place fish in a shallow metal, foil, or plastic pan. Cover the fish with water and freeze it. To prevent ice from evaporating while stored in freezer, wrap the pan in freezer paper after the water is frozen; label and store in freezer.

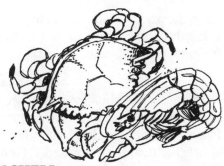

SEAFOOD RECIPES

CLAMS IN SHELL
See General Steps in Freezing (pp. 169-170). Place live clams in freezer bags and press out air. Fasten seals securely and freeze.

SHUCKED CLAMS
See General Steps in Freezing (pp. 169-170). Wash clams and shuck them. Clean the meat thoroughly and drain. Pack clams in containers, leaving a ½-inch headspace. Fasten lids or seals securely and freeze.

CRABS IN SHELL
Use fresh live crabs. See General Steps in Freezing (pp. 169-170). Drop crabs into pot of boiling water for 3 to 4 minutes (until dead). Remove and rinse. Remove internal organs and lungs. Separate back from claws and rinse. Pack backs and claws in separate containers, leaving no headspace. Do not add water. Fasten lids or seals securely and freeze. Use within 6 months.

SHELLED CRABS
Use fresh live crabs. See General Steps in Freezing (pp. 169-170). Drop crabs into pot of boiling water for 3 to 4 minutes (until dead). Remove and rinse. Remove internal organs and lungs. Remove meat from backs and claws. Pack crab meat in containers, leaving no headspace. Fasten lids or seals and freeze. Use within 3 months.

FISH ROE
See General Steps in Freezing (pp. 169-170). Thoroughly wash fish roe. Pack it in containers, leaving a ¼-inch headspace. Fasten lids or seals securely and freeze.

LOBSTERS IN SHELL
For best quality, lobster should be frozen uncooked. See General Steps in Freezing (pp. 169-170). Freeze the lobster whole. Or, clean the lobster and freeze just the shell portions that contain the edible meat. Some lobsters have large front claws that contain edible meat while others have edible meat mainly in the tail. Freeze lobster in the shell to avoid drying out the meat.

Wrap the whole lobster or lobster portions in moisture-vapor-resistant paper, using the drugstore wrap (p. 177) or the butcher wrap (p. 178). Seal it with freezer tape and freeze.

SHUCKED OYSTERS

Use fresh, good quality oysters. See General Steps in Freezing (pp. 169-170). Wash unshucked oysters. Steam them for 5 to 10 minutes to aid in shucking. Place a colander in a large bowl. Pour oysters and liquor into bowl as each is shucked. After all are shucked, lift out oysters only in the colander. Put them in a salt-water solution (2 tablespoons salt to each quart water) for a few minutes, stirring constantly. Drain for 5 minutes. Strain oyster liquor through a sieve. Pack oysters in containers and cover with the liquor, leaving a ½-inch headspace. Fasten lids securely and freeze. Use within 6 to 9 months.

SHUCKED SCALLOPS

Scallops should be alive until shucked. A live scallop will keep its shell tightly closed or will close it when tapped. See General Steps in Freezing (pp. 169-170). Shuck the scallops. Pack them in containers, leaving a ½-inch headspace. Fasten lids or seals securely and freeze.

COOKED SHRIMP

Use fresh, good quality shrimp. See General Steps in Freezing (pp. 169-170). Wash shrimp and remove head, feet, shell, and sand vein. Rinse. Drop shrimp in enough boiling water to completely cover them. Let water return to a boil. Remove from heat. Shrimp should turn pink within 3 to 5 minutes. Remove and rinse in ice water. Pack shrimp in containers, leaving a ¼-inch headspace. Fasten lids or seals securely and freeze. Use within 3 months.

PEELED SHRIMP

Use fresh, good quality shrimp. See General Steps in Freezing (pp. 169-170). Wash shrimp and remove head, feet, shell, and sand vein. Rinse. Pack shrimp in containers and partially cover with ice water, leaving a 1-inch headspace. As water freezes, it will expand, covering shrimp. Fasten lids securely and freeze. Use within 6 months.

SHRIMP IN SHELL

Use fresh, good quality shrimp. See General Steps in Freezing (pp. 169-170). Wash shrimp and remove head and feet. Cut the shell enough to remove the sand vein, but do not remove the shell. Rinse shrimp. Pack shrimp in containers and partially cover with ice water, leaving a 1-inch headspace. As water freezes, it will expand, covering shrimp. Fasten lids securely and freeze. Use within 6 months.

FREEZING EGGS

Selecting
Use only fresh, clean eggs. Dirt on the shells may cause spoilage. See General Steps in Freezing (pp. 169-170).

Preparing
Break each egg into a saucer and inspect it for freshness, blood spots, and pieces of shell before mixing with other eggs. If freezing yolks and whites separately, separate them into two saucers while breaking them.

Whole Eggs. Stir eggs slightly with fork, just enough to mix yolks and whites.

Egg Yolks. Beat yolks just a little, but not enough to make them fluffy. To each six egg yolks, add 1 teaspoon sugar or ½ teaspoon salt to prevent coagulation of the solids during storage.

Egg Whites. Whites may be frozen just as they are. Or, they may be gently mixed and strained through a sieve.

Packaging
Pack eggs in freezer containers, leaving a ½-inch headspace. Fasten lids securely and freeze. Use within 6 months.

Thawing and Using
Thaw only the amount you expect to use at one time. Thaw in refrigerator in the covered freezer container. Use thawed eggs promptly. Frozen eggs may be used in the same ways you use fresh eggs which have been stirred or mixed. Frozen egg whites are just as suitable for making angel food cakes as fresh whites.

Measuring
Use the following measures for substituting thawed eggs for fresh ones:
- 1 tablespoon yolk = 1 egg yolk
- 2 tablespoons white = 1 egg white
- 3 tablespoons mixed yolk and white = 1 egg

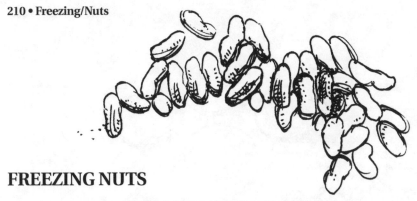

FREEZING NUTS

BOILED PEANUTS IN SHELLS

Use fully mature peanuts. See General Steps in Freezing (pp. 169-170). Wash shells with a mild detergent. Rinse well in clear water. Boil enough salt water (1 cup salt per gallon of water) in a large pot to cover peanuts. Add nuts to water, cover pot, and boil 45 minutes or until kernels are tender. Drain and cool. Pack peanuts in containers, leaving a ½-inch headspace. Fasten lids or seals securely and freeze.

Note: Peanuts may be boiled in salt water for only 30 minutes before freezing. When ready to use, remove nuts from freezer and partially thaw. Then, prepare another pot of boiling salt water and add peanuts. Boil for 15 minutes or until kernels are tender.

GREEN PEANUTS IN SHELLS

Use fully mature peanuts. See General Steps in Freezing (pp. 169-170). Wash shells with a mild detergent. Rinse well in clear water. Boil enough water to cover nuts. Add peanuts and blanch for 10 minutes. Drain well and cool. Pack peanuts in containers, leaving a ¼-inch headspace. Fasten lids or seals securely and freeze.

SHELLED PEANUTS

Use fully mature peanuts. See General Steps in Freezing (pp. 169-170). Wash shells to remove surface dirt. Shell peanuts. Pack nuts in freezer containers, leaving a ¼-inch headspace. Fasten lids or seals and freeze.

SHELLED PECANS

Use well-developed, fresh pecans. See General Steps in Freezing (pp. 169-170). Shell pecans and allow the nutmeats to dry. Remove any trash or bits of shell. As soon as they are dry, pack them in freezer containers, leaving a ¼-inch headspace. Fasten lids or seals securely and freeze. Use within 12 to 18 months.

FREEZING DAIRY FOODS

BUTTER

Use only high quality butter made from pasteurized cream. Unsalted butter is better for freezing than salted. See General Steps in Freezing (pp. 169-170). Wrap in freezer paper or place in containers. Fasten lids or seals securely and freeze.

CHEESE

Most cheeses become crumbly and mealy when frozen. Small pieces, no more than 1 pound or 1 inch thick, of the following varieties may be frozen satisfactorily: brick, Cheddar, Edam, Gouda, Muenster, Port du Salut, Swiss, provolone, mozzarella, Camembert, blue, Roquefort, and Gorgonzola. See General Steps in Freezing (pp. 169-170). Pack cheese in containers, leaving no headspace. Fasten lids or seals securely and freeze.

MILK

Use only homogenized milk. See General Steps in Freezing (pp. 169-170). Do not freeze milk in milk cartons or jugs. Pour it into appropriate moisture-vapor-resistant freezer containers, leaving a ½-inch headspace for pints; 1-inch for quarts. Fasten lids or seals securely and freeze. Milk may need to be put in the blender when thawed or stirred to return it to its natural consistency.

WHIPPING CREAM

Whipping cream separates when frozen and is not recommended for freezing. However, you can freeze whipped cream satisfactorily. See General Steps in Freezing (pp. 169-170). To freeze, whip the cream, adding sugar and vanilla. Drop whipped cream in small mounds on a baking sheet. Place it in the freezer. As soon as mounds are frozen, place them in freezer containers. Fasten lids or seals securely and return to freezer.

FREEZING COOKED FOODS

Freezing cooked foods saves time and labor in meal preparation. Entire meals can be prepared in advance; then, only a short preparation time is required the day they are served. Also, party refreshments may be prepared, packaged, and frozen in advance.

With a freezer at home, there is often a temptation to wrap and store leftovers in a haphazard fashion. These small packages can soon clutter up your freezer. The following tips will help you get the most efficient use of your equipment. Use all cooked foods within 3 months or less.

Breads

• Biscuits can be frozen baked or unbaked; however, the quality of unbaked biscuits is not generally as good as that of prebaked biscuits. Always use double-acting baking powder when making biscuits to freeze. Pack unbaked biscuits on a baking sheet and wrap in freezer paper. Use them within 2 to 3 weeks. Bake while still frozen.

• Biscuits are better if baked until they have risen and have just begun to turn brown before freezing. Cool biscuits, pack in freezer material, and freeze. Use within 3 months. When ready to use, do not thaw. Finish baking biscuits in a hot (425°F) oven until brown.

• Time is not saved by freezing unbaked muffins. Bake them until they have risen and have begun to turn brown; cool muffins, pack in freezer material, and freeze. When ready to use, do not thaw. Finish baking muffins in a hot (425°F) oven. Use frozen muffins within 3 months.

• You can freeze muffin batters. However, frozen batters will not rise as high, so always use double-acting baking powder. Place paper baking cups in muffin tins and fill two-thirds full. Package and freeze. Thaw before baking. Use frozen batters within 1 month.

• Cooked doughnuts, date and nut breads, and fruit breads are all good products for freezing. Cook until done; then cool. Package in freezer material. Use within 3 months.

• Use any standard roll recipes. Bake until they have fully risen and have begun to turn brown. Cool rolls. Package in freezer material and freeze. When

ready to use, place unthawed rolls in a hot (425°F) oven to finish baking. Use frozen rolls within 3 months.
• To freeze unbaked rolls, use a recipe that has been developed especially for freezing.

Desserts

• There is less chance for failure if cakes are baked before freezing. After baking, remove cake from pan and cool at room temperature. Put two layers of wax paper between each cake layer. Pack in freezer paper or a freezer bag. Freeze the cake and put it into a box to prevent mashing during storage.
• Any butter cake, angel cake, or sponge cake freezes well.
• Fruitcakes can be frozen and stored longer than other types of cake. The fruits blend and mellow during storage, and the cake stays moist because of the fruit.
• The best frostings and fillings for freezing are (1) uncooked frostings with confectioner's sugar and margarine or butter; and (2) cooked-candy type with honey or corn syrup as one of the ingredients.
• Most soft frostings, boiled icings, and cream fillings are not recommended for freezing.
• Lemon-cheese fillings may turn dark when frozen.
• To freeze a frosted cake, put the bottom layer of the baked cake on a piece of cardboard covered with freezer foil or freezer paper. Ice the cake, adding the top layer. Place iced cake in the coldest part of the freezer for 2 hours or until icing is firm. Then, wrap frozen cake in freezer material and store it in the freezer.
• A metal cake box with a tight-fitting cover can be used for storing a cake in the freezer.
• Thaw unfrosted, baked cakes in their freezer wrappings in the refrigerator or at room temperature. If unfrosted cake is wrapped in freezer foil, the ends can be loosened and the cake placed in a 200°F oven for 15 to 20 minutes until the cake is slightly warm. Then, it can be iced or eaten plain.
• To defrost an iced cake, remove it from the freezer and unwrap it. Place it in the refrigerator to thaw or let it thaw under a cake cover. The cake can be left wrapped for about 30 minutes at room temperature; then unwrap it to complete thawing.
• The most important factor in thawing iced cakes is to limit the amount of warm air which comes in contact with the frozen icing. Moisture may collect and cause icing to become soft. Use iced or plain cakes within 3 months.
• Mince pies, fruit pies, sweet potato pies, and pumpkin pies freeze satisfactorily unbaked. Compact pies or those with solid fillings, such as mince, pumpkin, and sweet potato, keep better over several months than some of the less compact, juicier fruit pies.

- Deep dish pies which have no undercrust to get soggy freeze well.
- When making a fruit pie to freeze, add cornstarch to the fruit juice and cook until clear. Cool and add the thickened juice to the fruit. Do not cut vents in crust before freezing, but be sure to cut vents before baking. Peaches and apples discolor readily; pretreat with ascorbic acid, citric acid, or lemon juice.
- Package pies in the pie plate. Freeze them and put them in a box to protect the upper crust. Wrap box in freezer paper and freeze.
- To cook frozen unbaked pies, remove from freezer, slit crust, if necessary, and place frozen pie in a hot (450°F) oven for about 10 minutes. Then, bake at 350°F for about 50 minutes or until done.
- Custard pies with wheat flour as a thickening agent do not freeze well either before or after baking because the filling becomes grainy and separates.
- Chocolate and lemon chiffon pies as well as pecan pies should be baked before freezing.
- Meringue becomes tough and shrinks when frozen. Freeze unbeaten egg whites in another package and thaw when ready to make meringue. Or, add whipped cream toppings just before serving.
- Crumb crust pastry and pastry shells, either baked or unbaked, can be frozen satisfactorily.
- Use all frozen pies within 3 months.
- Prepare baked apples by your favorite recipe. Then, cool apples quickly, pack in freezer containers, and freeze. Frozen baked apples are not so attractive in appearance as freshly baked ones. Thaw baked apples in the refrigerator or at room temperature in the original wrappings, unless you use the oven for thawing.
- Almost all types of cookies can be frozen. After baked cookies are cool, package and freeze. Frozen unbaked cookies do not rise so much and, in most cases, must be thawed before baking. Refrigerator cookies can be baked without thawing.

Meat and Poultry

- If frozen cooked meats, such as roast beef, ham, and poultry, are overexposed to air during storage, they develop warmed-over flavors and odors. To prevent this, freeze meats in large pieces or in very solid packs. If small pieces of meat are to be frozen, first cover them with gravy or sauce.
- Always remove excess fat on any meat or poultry that is to be frozen.
- To freeze chicken or turkey and dressing, remove bones before storing to save freezer storage space. Pack meat tightly to eliminate as many air pockets as possible. Package the dressing separately from the meat.

- To cook chicken for freezing, simmer it in lightly salted water until tender; chill and remove meat from bone. Cut the meat into 1-inch pieces. Pack meat tightly and wrap in freezer paper. Chicken or turkey prepared this way can be creamed or used in pies, salads, or loaves.
- Meat and poultry stews, pies, and combination dishes freeze well. Use your favorite recipe, but do not use any vegetable that does not freeze well after it is cooked. For example, white potatoes tend to become soft, and hard-cooked egg whites should be mashed through a sieve before freezing.
- When reheating completely cooked meat, heat only long enough to get thoroughly hot; otherwise, it will have a warmed-over flavor.

Salads

- Freezing destroys the crispness of raw vegetables. They lose their color and flavor and, therefore, are not suitable for use in frozen salads.
- Cooked meat and poultry and some canned and fresh fruits make delicious frozen salads.
- The most satisfactory frozen salad is made with a gelatin base and a choice of cream cheese, cottage cheese, whipped cream, sour cream, or mayonnaise and frozen in refrigerator trays.

Sandwiches

- Any variety of bread can be used for sandwiches that will be frozen. Day-old bread is better than fresh.
- Sandwich fillings suitable for freezing may contain cheese, sliced or ground meat, sliced or ground poultry, or cooked egg yolks.
- Fillings not suitable for freezing are those that contain raw vegetables, whites of hard-cooked eggs (unless mashed through a sieve), and jelly.
- The following spreads or fillings make good frozen sandwiches: dried beef and cheese; cream cheese, honey, and orange rind; minced egg and ham; meat and olives; cream cheese and finely chopped ginger; ham salad; peanut butter; cream cheese, olives, and nuts; cream cheese and nuts.
- Don't use too much mayonnaise in sandwich spreads because it separates when freezing and soaks into the bread.
- Use softened butter to spread on bread before putting on filling.
- Wrap sandwiches in freezer material and place in a box to keep them from being mashed. Keep sandwiches covered while thawing in the refrigerator.

Soups

- Frozen soups taste fresher than soups preserved in other ways. Use your favorite recipe and cook to the proper consistency. Cool quickly and remove excess fat. Pack in freezer containers and freeze.
- Since soups contain a large amount of water, make them thick for freezing and then thin them with water or milk at the time they are heated.

Vegetables

• Most completely seasoned and cooked vegetables have a warmed-over flavor when frozen. If cooked vegetables are to be frozen, cook until about three-fourths done; then, cool quickly. Package in freezer containers and freeze. When reheating, they will develop their characteristic aroma.

• Sweet potatoes may be baked, cooled, and sealed in freezer containers. Freeze immediately and store. Use within 3 months. Cooked pureed sweet potatoes also freeze well.

• To freeze Irish potatoes, use good quality baking potatoes. Bake and, when done, cut potatoes in half, lengthwise. Scoop potato from skins, forming shells. Mash potatoes with seasoning and milk. Restuff mashed potatoes in shells. Pack in freezer material and freeze. Use within 3 months. To reheat, place thawed or frozen potato halves in a moderate (350°F) oven. Heat until hot.

• To freeze french fries, peel and slice potatoes into ⅜-inch strips, 2 to 3 inches long. Blanch for 2 minutes in lightly salted boiling water to help prevent discoloring. Drain well and pat dry. Fry quickly to a light brown. Drain well and cool. Do not add salt. Pack in freezer containers and freeze. Use within 4 weeks. To serve, spread potatoes on a baking sheet. Heat them to finish browning in a hot oven (400°F) about 5 to 6 minutes. Serve at once.

• Another method for freezing french fries is to rinse potato strips in cold water to remove starch. Dry thoroughly on absorbent paper towels. Arrange potatoes one layer deep in a shallow baking pan. Brush all sides of potatoes with melted butter or margarine. Bake in a hot (450°F) oven until they begin to turn brown, turning occasionally. Remove from oven and put in cool baking pan in refrigerator to cool quickly. Pack in freezer containers and freeze. Use within 1 month. To serve, return frozen fries to shallow baking ban and bake in 450°F oven until thawed, tender, and golden brown. Season to taste.

Drying

Of all the methods of preserving food, drying is the simplest. Dried foods are preserved by removing the water that is in them, thus preventing the growth of bacteria, yeasts, and molds and slowing down the action of enzymes. When you are ready to use the food, the water is restored by soaking.

Table 18. Pounds and Pints of Dried Foods From Fresh

Type of Food	Pounds of Fresh Food	Amount After Drying	
		Pounds	Pints
Apples	12	1¼	3
Lima Beans	7	1¼	2
Snap Beans	6	½	2½
Beets	15	1½	3 - 5
Broccoli	12	1¼ - 1½	12 - 15
Carrots	15	1¼	2 - 4
Celery	12	¾	3½ - 4
Corn	18	2½	4 - 4½
Greens	3	¼	5½
Onions	12	1½	11½ sliced 4½ chopped
Peaches	12	1 - 1½	2 - 3
Pears	14	1½	3 quartered
Peas	8	¾	1
Pumpkin	11	¾	3½
Squash	10	¾	5
Tomatoes	14	½	2½ - 3

Drying requires very little equipment and is relatively inexpensive. Dried, or dehydrated, foods weigh less and take less storage space than most fresh, canned, or frozen foods, depending on the packaging material used (see Table 18, above). Also, the food tissue is not damaged during drying as it often is when canning or freezing.

While drying is not complicated, it does take a long time and a lot of attention. The drying trays must be shifted and the food rearranged on the trays frequently for even drying. Also, once the drying process has started, it must be continued without interruption.

FOOD SAFETY AND QUALITY

Not all foods are suited for drying. Vegetables and fruits can be dried safely. Seeds and herbs make good dried products. But most vegetables and some fruits are better preserved by freezing or canning. Meat jerky has long been a favorite dried meat, but the danger of *E. coli* bacteria has made the safety of drying meat at home questionable.

Food Safety

The best temperature for drying food is 140°F. If higher temperatures are used, the food will cook on the outside, causing "case hardening." When that happens, the internal moisture cannot escape and the food will eventually mold. The drying temperature should never be raised above 140°F to speed the process or for any other reason. Meat must be heated to an internal temperature of 180°F to kill *E. coli* bacteria.

Food Quality

In order to preserve as much food value, natural flavor, and cooking qualities as possible, several procedures are involved in the drying process.

• The food must first be prepared for drying just as it is for other forms of preservation: thoroughly washing, peeling, coring, and cutting into the appropriate size pieces.

• The food must be pretreated to stop enzyme activity that can cause spoilage and affect appearance and taste.

• The food is then dried to stop the growth of bacteria.

• Foods that are sun dried and beans that are vine dried must be pasteurized to kill any insects and insect eggs that get on the food during the drying process.

• Fruits, which retain 20 percent of their moisture after drying, must be conditioned to distribute the moisture evenly throughout each piece.

GENERAL STEPS IN DRYING

1. Assemble or make ready the drying equipment (see Dryers, pp. 219-221) that you will use. If you will sun dry your food, be sure the weather forecast predicts several consecutive hot, dry, breezy days (see Sun Drying, p. 220).

2. Assemble storage containers (see Containers, p. 222). Wash in hot soapy water and rinse in hot water. Drain and allow them to air dry.

3. Select good quality, fresh, mature food that will make a quality dried product (see Selecting, pp. 222-223; also see Table 19, p. 222, and Table 20, p. 223). Discard any that are wilted, moldy, decayed, or bruised.

4. Wash vegetables and fruits thoroughly until every trace of soil is removed (see Preparing, pp. 223-224). Trim away any small damaged or diseased spots from otherwise healthy food.

5. Trim, peel, cut, slice, or shred vegetables and fruits into uniform-size pieces, according to instructions for the particular food (see Recipes, pp. 230-236).

6. Pretreat the food by the method that is recommended or that you prefer (see Pretreating, pp. 224-225; also see recipes for specific foods, pp. 230-236). Then, cool the pieces and allow them to dry.

7a. For sun drying (see How to Sun Dry, p. 226), place cheesecloth or netting on drying rack. Then, place treated food on dryer trays (see Trays, p. 221), cover with cheesecloth or net, and place on drying rack in the sun.

b. For drying in the oven, see How to Oven Dry (pp. 226-227); for drying in a dehydrator, see manufacturer's instructions or How to Dehydrator Dry (p. 227). For drying herbs, hot peppers, or nuts, see Air Drying (p. 221).

8. After food has dried the specified time (see Tables 26, p. 249, and 27, p. 250, or see recipe for specific food, pp. 230-236), check to see if it is dry enough (see Testing for Dryness, pp. 227-228). If not, replace it on the trays and continue to dry until it is done.

9. If food was sun dried or air dried, pasteurize it (see Pasteurizing, p. 228).

10. If food is a fruit, condition it (see Conditioning Fruits, p. 228).

11. Pack food in clean containers (see Packaging, pp. 228-229) and label them.

12. Store containers in a cool, dark, dry place (see Storing, p. 229).

To use dried foods, see Reconstituting Dried Foods (pp. 229-230).

MATERIALS AND EQUIPMENT

Needed:
- Drying equipment (see Dryers, pp. 219-221)
- Drying trays (see Trays, p. 221)
- Blanching equipment
- Storage containers
- Oven thermometer
- Timer or clock
- Measuring cup
- Sharp knife
- Clean cloths
- Hot pads

Dryers

The kind of equipment you need for drying depends on the method you use. Three important factors are needed to dry food: heat, low humidity, and air movement. The heat does the drying, but low humidity aids the drying process. Food contains a lot of water. To dry food, the water must move from the food to the surrounding air. If the surrounding air already contains a lot of moisture, then drying will be slowed down.

Increasing the air movement will speed up the drying process. By moving the air that has absorbed moisture from the food, drier air will take its place, ready to receive more moisture.

Sun Drying. If you can depend on having several hot, dry, breezy days in a row, this method can be used to dry fruits outside. Fruits are high in acid and sugar, making them safe for sun drying. Vegetables can be dried in the sun, but they are more at risk for food spoilage because they are low-acid foods. A quicker, more controlled drying method is safer for vegetables (other than beans).

Make a dryer by placing a clean window screen or a wooden rack on concrete blocks. Locate the dryer on a paved driveway or patio because the bare ground may be moist, making the drying process more difficult. If you cannot use a paved area, place a sheet of aluminum or tin under the screen or rack. The reflection of the sun on the metal will increase the drying temperature. Screens that are safe to use are stainless steel, teflon-coated fiberglass, and plastic.

Caution: Do not use galvanized screen. The zinc or cadmium coating can cause a dangerous reaction when it comes in contact with acid foods. Also, avoid uncoated fiberglass, vinyl, aluminum, and copper screening. Copper destroys vitamin C and increases oxidation. Aluminum tends to discolor and corrode.

Solar Drying. Efforts to improve sun drying led to solar drying. This method uses the sun as the heat source, but a specially designed dehydrator increases the temperature and air current to speed up the drying time. Shorter drying time reduces the risk of spoilage or molding. Solar dryers may need turning or tilting throughout the day to capture the direct, full sun. There are several models available, or you can make a solar drying unit at home.

Oven Drying. This is a safer method of preservation for both fruits and vegetables than sun or air drying. Foods dried in the controlled heat of an oven will generally be superior to sun-dried foods in color, flavor, cooking quality, and nutritive value. Also, oven-dried foods are more likely to be free from insects and insect eggs that often appear in foods dried outside or in an open room. Controlled-heat drying has the advantage of reducing total drying time because it can continue after sundown and during rainy days. Oven drying is sometimes slower than dehydrators, though, because conventional ovens do not have a built-in fan for air movement. Some types of convection ovens do have a fan. Oven drying also has the disadvantage of tying up your oven for up to 5 or 6 hours at a time.

Dehydrator Drying. Using a dehydrator is probably the most efficient way to dry foods. It has all the advantages of oven drying without the disadvantages. A food dehydrator is relatively small but, in many cases, dries food in much less time than the other methods. It has a heating element and a fan and vents for air circulation. A dehydrator may cost from $50 to $400, depending on size and features. Plans are available for building a dehydrator. Call your county Extension office for information. Homemade dehydrators can save money but are not as efficient as those you can buy.

Vine Drying. This is a method of drying beans (navy, kidney, butter, great northern, lima, lentils, and soy). They are simply left on the vine until the beans are dry enough to rattle inside their pods. When the vines and pods are dry and shriveled, the beans are ready to be picked and shelled. No pretreatment is necessary.

Air Drying. Also called room drying, this method can be done in a clean, well-ventilated room, attic, car, camper, or screened porch. Herbs, hot peppers, nuts in the shell, and partially dried foods that had been sun dried are the most common air-dried items. Herbs and peppers can be strung on a string or tied in bundles and suspended from overhead racks in the air until dry. Enclosing them in paper bags, with openings for air circulation, protects them from dust, loose insulation, and other pollutants. Nuts are spread in single layers on papers. Partially sun-dried foods should be left on their drying trays. See recipes (pp. 242-245) for more information about specific foods.

Trays

Most methods of drying require that the food be spread in a thin layer on trays. Dehydrators and solar dryers that you buy will include trays. For sun drying, trays with sides that are higher than the level of the food are better than flat trays because:
- the sides keep food from sliding off when the trays are moved;
- they make stacking easier for carrying inside at night;
- the sides keep other trays from resting on the food when they're stacked;
- they provide an edge to which a protective covering may be fastened.

Trays that will be used outside should be a size that's easy to handle. Trays used in an oven or a homemade dehydrator should be at least 1½ inches shorter in length and width than the oven or dehydrator to allow good air circulation.

Blanchers

A blancher is needed for pretreating vegetables (see Figure 12, p. 175). A steamer can be used for steam blanching. If you do not have a blancher or steamer, you can use a large pot with a tight-fitting lid and a wire basket, colander, or sieve that will fit inside to hold the food. See Blanching (p. 224) for more information about needs for steaming and water blanching.

Containers

Glass jars and metal cans or boxes with tight-fitting lids are good containers for storing dried foods. Moisture-vapor-resistant plastic freezer containers are also good for dried foods. Heavy-duty self-sealing plastic bags can be used, but they are susceptible to insects and mice.

METHODS AND PROCEDURES

Selecting

Foods selected for drying should be of the highest quality possible—fresh and fully ripened. Wilted or inferior fruits and vegetables will not make a good dried

Table 19. Suitability of Vegetables for Drying

Vegetable	Suitability
Asparagus	Poor to fair
Lima Beans	Fair
Snap Beans	Fair to good
Beets	Fair to good
Broccoli	Not recommended
Brussels Sprouts	Poor
Cabbage	Fair
Carrots	Good
Cauliflower	Poor
Celery	Poor
Sweet Corn	Good
Popcorn	Good
Cucumbers	Poor
Eggplant	Poor to fair
Garlic	Good
Greens	Poor
Horseradish	Good
Lettuce	Not recommended
Mushrooms	Good
Okra	Fair to good
Onions	Good to excellent
Parsley	Good
Garden Peas	Fair to good
Hot Peppers	Excellent
Sweet Peppers	Good
Irish Potatoes	Good
Pumpkins	Fair to good
Radishes	Not recommended
Rutabagas	Fair to good
Summer Squash	Poor to fair
Winter Squash	Not recommended
Sweet Potatoes	Fair
Tomatoes	Fair to good
Turnip Roots	Fair to good
Zucchini	Poor to fair

Table 20. Suitability of Fruits for Drying

Fruit	Suitability
Apples	Excellent
Apricots	Excellent
Bananas	Good
Blackberries	Not recommended
Cherries	Excellent
Citrus Fruits	Not recommended
Citrus Peel	Excellent
Coconuts	Excellent
Crab Apples	Not recommended
Cranberries	Poor
Figs	Excellent
Grapes	Excellent
Melons	Poor
Nectarines	Excellent
Peaches	Excellent
Pears	Excellent
Persimmons	Fair
Pineapples	Excellent
Plums	Good
Prune Plums	Excellent
Rhubarb	Good
Strawberries	Fair to good

product. Foods that are immature lack flavor and color. Those that are overmature may be tough and fibrous or soft and mushy.

Preparing

Prepare foods for drying immediately after gathering. Then, begin drying as soon as they are prepared. Sort and discard any defective food. Decay, bruises, or mold on any piece may affect an entire batch of food being dried. Wash or clean all fresh fruits and vegetables thoroughly to remove any dirt or chemical residues.

Fruits. Core or pit fruits and cut them in halves or slices, depending on the recipe you are using (see pp. 230-232). They should be uniform in size to dry more evenly. Thin, peeled slices dry more quickly. The peel may be left on, but it takes longer to dry. Some fruits may be dried whole, but whole fruits take the longest to dry. Whole fruits should be dried in a dehydrator, especially on days with high humidity. They should be "checked" or cracked to reduce drying time. To "check" a whole fruit, place it in briskly boiling water and then, immediately, in very cold water.

Vegetables. Trim, peel, cut, slice, or shred vegetables, depending on the recipe you are using (see pp. 233-236). Keep pieces uniform in size so they will dry at the same rate. Remove any fibrous or woody portions and core those necessary. Remove any small decayed and bruised areas on otherwise quality vegetables. Prepare only as many vegetables as can be dried at one time. Prepared vegetables, even in the refrigerator, will lose quality and nutrients.

Pretreating

Vegetables and fruits must be pretreated to slow down or stop enzyme activity. Enzymes are responsible for color and flavor changes during ripening. These changes will continue during drying and storage unless the produce is pretreated. There are several ways to pretreat, some depending on the type of produce and some depending on your preference.

Blanching. This is the recommended pretreatment for most vegetables. Blanching before drying sets the color; it speeds drying by softening the tissues, which hastens the escape of moisture; it helps prevent undesirable flavor changes during storage; and it helps ensure satisfactory reconstitution during cooking. Vegetables may be blanched by steaming or boiling. The steam method is preferred because it does not destroy as many water-soluble nutrients.

Fruits can be steam blanched if they will be dried in the oven or a dehydrator. Blanching will give fruits a darker color than other pretreatment methods. It may also give a slightly cooked flavor to fruits such as apricots, peaches, and pears. Blanched fruits may be soft and a little difficult to handle.

Steam Blanching. To steam blanch, use a steamer or a deep pot with a tight-fitting lid and a wire basket, colander, or sieve that will fit inside. There should be about 5 inches from the bottom of the basket to the bottom of the pot. Pour 2 inches of water into the pot and bring it to a boil. Loosely place the vegetable pieces in the basket no more than 2 inches deep. Place the basket into the pot, making sure the boiling water does not come in contact with the vegetables. Cover the pot and steam the vegetables until each piece is heated through (see Table 27, pp. 250-251). Test by cutting through a piece to see if it is translucent nearly to the center. Remove basket of vegetables and dip immediately into cold water, just long enough to stop the cooking action. When they feel only slightly hot to the touch, drain by pouring vegetables directly onto the drying tray held over the sink.

Water Blanching. To blanch with boiling water, use a blancher or a deep pot with a tight-fitting lid. Add enough water to cover a small amount of vegetables and bring to a boil. Use only a small amount of vegetables at a time so that adding them won't stop water from boiling. Add vegetables to the boiling water and cover pot. Boil for the recommended time (see Table 27, pp. 250-251). Remove vegetables but save water and reuse for next batch. Add fresh water when needed. Place blanched vegetables immediately into cold water, just long enough to stop the cooking action. When they feel only slightly hot to the touch, drain by pouring vegetables directly onto the drying tray held over the sink.

Ascorbic Acid. Ascorbic acid can be bought in three forms. Pure ascorbic acid (vitamin C) is an antioxidant that helps keep fruit from darkening as it is being prepared for drying. For most fruits, 1 teaspoon **ascorbic acid crystals** per cup of water is enough. If the fruit is a variety that darkens badly, increase the amount of acid. One cup of **ascorbic acid mixture** should treat about 5 quarts of prepared fruit. Sprinkle the mixture over the fruit pieces, turning the fruit gently to treat each piece. Or, use 1 teaspoon of **ascorbic acid powder** in 2 cups of water and place the prepared fruit in the solution for 3 to 5 minutes. Remove the fruit, drain well, and place on the drying tray.

Sulfite Dip. Sodium bisulfite, sodium sulfite, or sodium metabisulfite can be used for pretreating fruits. Look for these products at drugstores or hobby shops where wine-making supplies are sold. Dissolve the sulfite product in water (about ¾ to 1½ teaspoons bisulfite per quart of water; 1½ to 3 teaspoons for sulfite; 1 to 2 teaspoons for metabisulfite) and soak fruit in mixture 5 minutes for slices and 15 minutes for halves. Remove and rinse fruit lightly under cold water; then, place on the drying tray. Mix fresh solution for each batch of fruit. Solution can be used only once.

Caution: Sulfite residue in foods have caused serious reactions in some people who have asthma. Do not serve foods treated with a sulfite dip to any person who has asthma. Also, do not use bisulfate. Be sure to use U.S.P. (food grade) or Reagent Grade (pure) sulfite products. Practical Grade is not pure.

Saline Solution. Fruit may be soaked in a solution of 2 to 4 tablespoons of salt dissolved in 1 gallon of water. Soak fruit pieces for 10 minutes, drain, and place on the drying tray.

Syrup Blanching. Fruit blanched in syrup keeps its color fairly well, but the dried fruit will be similar to candied fruit. The fruit will take longer to dry and will attract insects if dried outside. To make the syrup, mix 1 cup sugar, 1 cup light corn syrup, and 2 cups water in a saucepan. Bring syrup to a boil. Add 1 pound of prepared fruit and simmer 10 minutes. Remove pan from heat and let fruit sit in hot syrup for 30 minutes. Then, remove fruit, rinse lightly in cold water, drain on paper towels, and place on the drying tray.

Fruit Juice Dip. Fruit juices that are high in vitamin C—orange, lemon, pineapple, grape, and cranberry—can be used as a pretreatment, but they are not as effective as pure ascorbic acid. Also, the juice adds its own color and flavor to the fruit. Add prepared fruit pieces to juice in a bowl and soak 3 to 5 minutes. Remove fruit, drain well, and place on the drying tray.

Honey Dip. Many commercially dried fruits have been dipped in a honey solution. You can make a similar dip, but honey-dipped fruit is much higher in calories. To make the dip, add ½ cup sugar to 1½ cups boiling water and mix to dissolve sugar. Cool solution to lukewarm and add ½ cup honey. Place fruit pieces in dip and let them sit for 3 to 5 minutes. Remove pieces, drain well, and arrange on the drying tray.

226 • Drying/General

Drying

How to Sun Dry. It will take several days to dry foods outside in the sun. Prepare the fruit or vegetable and have it ready to start drying by 9:00 a.m. The temperature needs to reach 90° to 100°F by noon, and the humidity should be less than 60 percent. Select a drying area that offers some protection from dust, dirt, and animals. An electric fan directed over the food will hasten drying and help keep insects away. Read General Steps in Drying (pp. 218-219).

1. Place pretreated food pieces one layer deep on a tray that has been covered with a clean, dry cheesecloth or netting.
2. Cover food with a thin cloth, being sure cloth does not touch the food. Weight cloth so that insects cannot get under it to lay eggs on the food.
3. Turn pieces over every 2 hours or so to speed drying.
4. Bring the trays in at night.
5. Put trays out the next day. Continue to do this until food is dry. It may take 3 or 4 days, depending on type of food, size, and air temperature. The food will be leathery when drying is complete.

If the humidity becomes higher than 60 percent or if it should rain, finish drying food in the oven.

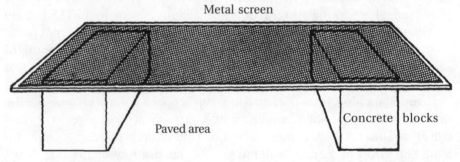

Figure 15. Outdoor Dryer

How to Oven Dry. To dry foods in your oven, be sure you can maintain a constant oven temperature of 140°F with the door propped open 4 inches and a fan directed into the oven. If you cannot, you should consider using another method.

1. Prepare no more than 6 pounds of fresh food for one oven load.
2. Preheat the oven at its lowest setting. Place an oven thermometer inside the oven to check the temperature throughout the drying process.
3. Prop the oven door open at least 4 inches for good air circulation. Place a fan outside the oven door so that it is directed through the opening and across the oven. Because the door is left open, the temperature may vary. The open door and the use of the fan may lower the temperature in your oven to

140°F even if there is no setting that low. Adjust the temperature dial as needed to keep the oven at 140°F.

4. Place the trays in the oven. They should be at least 1½ inches shorter than the width and depth of the oven. Leave about 2½ vertical inches between trays and allow at least a 3-inch clearance from the top and bottom of the oven.

5. Turn the pieces of food over, move the center pieces to the outside edges, rotate the trays from shelf to shelf, and turn the trays from front to back every half hour so that all pieces will dry more evenly. It is helpful to number the trays and keep a record of the changes.

6. It is easy to scorch food that overheats near the end of the drying time (see Tables 26, p. 249, and 27, pp. 250-251). If some pieces around the tray edges dry first, remove them. When drying time is almost complete, turn off the heat, open the door wide, and leave the trays in the oven until the remaining pieces are completely dry.

How to Dehydrator Dry. Follow manufacturer's directions for using a dehydrator. Generally, food is prepared and loaded on trays as for oven drying. Trays must be checked and rotated occasionally.

Testing for Dryness

Foods should be dry enough to prevent microbial growth and spoilage. Vegetables must dry out to 5-percent or less moisture content to be adequately preserved for storage. Most dried fruits are safely

Figure 16. Oven Drying

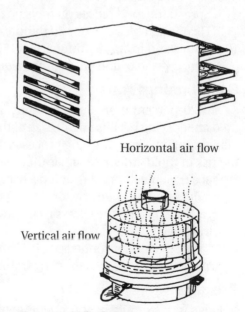

Horizontal air flow

Vertical air flow

Figure 17. Dehydrators

preserved when they still contain about 20 percent of their moisture. Berries, however, should be dried until they rattle when shaken together.

Vegetables. Dried vegetables should be brittle or crisp. Some would actually shatter if hit with a hammer. Because they are so dry, they do not need conditioning as fruits do. However, if they are sun or air dried, they will need to be pasteurized (see below).

Fruits. Dried fruits are often eaten without being rehydrated, so they should not be dehydrated to the point of brittleness. Dried fruits should be leathery and pliable. To test for doneness, cut several cooled pieces in half. There should be no visible moisture, and you should not be able to squeeze any moisture from the fruit. Also, fruits should not be sticky or tacky. If a piece is folded in half, it should not stick to itself. If fruits are sun or air dried, they will need to be pasteurized (see below). After sun/air-dried fruits are pasteurized or when fruits dried by other methods are done, they need to be conditioned (see below).

Pasteurizing

Any food dried in the sun or open air should be heated before packaging to destroy insects and insect eggs. To pasteurize:

1. Spread the food pieces on trays no more than 1 inch deep.
2. Heat vegetables or fruits 30 minutes in a 150°F oven; or, heat vegetables 10 minutes and fruits 15 minutes in a 160°F oven.
3. Cool vegetables and package immediately. Cool and then condition fruits.

Conditioning Fruits

The 20-percent moisture remaining in fruits is not always evenly distributed after drying because of the size of the piece or its location in the oven or dehydrator. The conditioning process will equalize the moisture and reduce the risk of mold growth on areas that may still be too moist.

Place the cooled dried fruit loosely in plastic or glass containers. Seal the containers and let them sit for 7 to 10 days. Shake the containers daily to separate the pieces. The excess moisture in some pieces will be absorbed by the drier pieces.

Also, check daily for condensation. If condensation develops in the container, return the fruit to the oven or dehydrator for more drying.

Packaging

Dried foods should be thoroughly cooled before packaging. Fill containers as tightly as possible without crushing the food. Use clean, dry, insect-proof containers. Glass jars, metal canisters or boxes with tight-fitting lids, and plastic freezer cartons are all good containers.

Pack foods in the amounts that they will most likely be used. Every time a package is opened, the food is exposed to air and moisture which can lower its quality.

Label containers with the name of the food and the date it was packaged.

Storing

Store dried foods in a cool, dry, dark place. Heat affects food quality. If dried food is kept cool, it keeps longer. Dried fruits can usually be stored for 1 year if kept at 60°F; 6 months at 80°F. Vegetables have about half the shelf-life of fruits.

Light causes dried foods to fade or deteriorate. If it is packed in glass or some other clear container, wrap the container with foil or paper to block out the light.

Also, moisture in the storage area can be absorbed by the dried foods. Check them frequently during storage to see that they are still dry. Foods that are affected by moisture, but are not yet spoiled, should be used immediately or redried.

Caution: Discard any dried foods that develop mold.

Reconstituting

The time it takes to reconstitute foods (restore the moisture) depends on the kind, the size of the pieces, and whether you use hot or cold water for soaking. Oversoaking will produce loss of flavor, and the food may get mushy or waterlogged. Also, drying does not destroy the bacteria, yeasts, or molds that may have been present on the foods. When the moisture is restored, they can again start to multiply. If soaking continues too long, spoilage will occur. Foods should be refrigerated if soaking will take longer than 1 hour.

Vegetables. One cup of dried vegetables makes about 2 cups of reconstituted vegetable. Cover the dried vegetables with cold water and soak until they have returned to their natural size and shape. This can take anywhere from 20 minutes to 2 hours, depending on the type and size of the pieces.

To cook reconstituted vegetables, heat them to a boil in the soak water to retain as much of the nutritive value as possible. Reduce heat and simmer until done. Allow the excess water to evaporate during cooking. Cooking time will range from 10 to 30 minutes, depending on the vegetable. Like fresh vegetables, dried vegetables lose flavor and texture if they are overcooked.

You can add dried vegetables that are cut into small pieces to soup, stew, or casserole recipes without reconstituting them.

Fruits. One cup of dried fruit makes about 1½ cups of reconstituted fruit. Add water to just cover the fruit; more can be added later, if needed. It takes up to 8 hours to reconstitute most fruits. Allow fruits to reconstitute before cooking.

To cook reconstituted fruits, cover and simmer in the soak water to retain nutrients. Since the drying process may change some of the starch in fruits to sugar, additional sweetening may not be needed. Adding a few grains of salt will help bring out the natural sweetness of most fruits. If a sweeter fruit is desired, add sugar or honey near the end of the cooking time. Lemon, orange, or grapefruit juice added just before serving helps give a fresh fruit flavor and adds vitamin C to the fruit.

FRUIT RECIPES

APPLE SLICES OR RINGS

See General Steps in Drying (pp. 218-219). Wash and peel apples; remove cores and cut into slices or rings about ⅛ inch thick. Steam blanch 3 to 5 minutes or syrup blanch 10 minutes. Or, you can use an ascorbic acid, fruit juice, or sulfite dip. Drain well.

Sun dry, 3 to 4 days

Dehydrator, 6 to 12 hours

APRICOT HALVES OR SLICES

See General Steps in Drying (pp. 218-219). Wash apricots and remove pits; cut in halves or slices, if desired. Steam blanch 3 to 4 minutes, water blanch 4 to 5 minutes, or syrup blanch 10 minutes. Or, you can use an ascorbic acid, fruit juice, or sulfite dip. Drain well.

Sun dry, 2 to 3 days

Dehydrator, 24 to 36 hours

BANANA SLICES

Use solid yellow or slightly brown-flecked bananas. Avoid bruised or overripe ones. See General Steps in Drying (pp. 218-219). Peel and slice bananas crosswise into ¼- to ⅜-inch pieces. Use a honey dip. Or, you can use an ascorbic acid, fruit juice, or sulfite dip. Drain well.

Dehydrator, 8 to 10 hours

BERRIES

See General Steps in Drying (pp. 218-219). Wash berries and drain; remove the caps. Dip berries into boiling water for 15 to 30 seconds to "check" skins. Then, drop them in ice water to stop cooking action. **Do not "check" strawberries or boysenberries.**

Dehydrator, 24 to 36 hours

CHERRIES

Use fully ripe cherries. See General Steps in Drying (pp. 218-219). Wash cherries, remove stems, and drain. Remove seeds. Cut in half, chop, or leave whole. Syrup blanch sour cherries for 10 minutes. Or, dip whole cherries in boiling water 30 seconds to "check" skins. Then, drop them in ice water to stop cooking action. No pretreatment is necessary for cut or pitted cherries.

Dehydrator, 24 to 36 hours

CITRUS PEEL

Use peels of citron, grapefruit, kumquat, lime, lemon, tangelo, and tangerine. Thick-skinned navel orange peel dries better than thin-skinned varieties. See General Steps in Drying (pp. 218-219). Wash fruit thoroughly. Remove outer 1/16- to 1/8-inch of peel. Remove the bitter white pith. No pretreatment is needed.

Dehydrator, 8 to 12 hours

FIGS

Use fully ripe figs. Immature fruit may sour before drying. See General Steps in Drying (pp. 218-219). Wash or clean whole figs with a damp cloth. Leave small figs whole; cut larger ones in half. Dip whole figs in boiling water for 30 seconds or more to "check" skins. Then, drop in ice water to stop cooking action. Drain on paper towels. Cut figs need no pretreatment.

Sun dry, 4 to 5 days
Dehydrator, 12 to 20 hours

GRAPES

See General Steps in Drying (pp. 218-219). Wash grapes and drain thoroughly. Leave seedless grapes whole. Cut seeded grapes in half and remove seeds. Dip whole grapes in boiling water for 30 seconds or more to "check" skins. Then, drop in ice water to stop cooking action. Drain on paper towels. Grape halves need no pretreatment.

Sun dry, 3 to 5 days
Dehydrator, 12 to 20 hours

NECTARINE QUARTERS OR SLICES

See General Steps in Drying (pp. 218-219). Wash nectarines and drain. Remove skins, if desired, by dipping in boiling water 30 to 60 seconds and then dipping into cold water to loosen skins. If steam or syrup blanching, leave nectarines whole; then halve, remove seeds, and quarter or slice after blanching. Steam or water blanch 8 minutes or syrup blanch 10 minutes. Or, you can use an ascorbic acid, fruit juice, or sulfite dip. Drain well.

Sun dry, 3 to 5 days
Dehydrator, 36 to 48 hours

PEACH QUARTERS OR SLICES

See General Steps in Drying (pp. 218-219). Wash peaches and drain. Remove skins, if desired, by dipping in boiling water 30 to 60 seconds and then dipping into cold water to loosen skins. If steam or syrup blanching, leave peaches whole; then halve, remove pits, and quarter or slice after blanching. Steam or water blanch 8 minutes or syrup blanch 10 minutes. Or, you can use an ascorbic acid, fruit juice, or sulfite dip. Drain well.

Sun dry, 3 to 5 days
Dehydrator, 36 to 48 hours

PEAR QUARTERS OR SLICES

See General Steps in Drying (pp. 218-219). Wash pears and drain. Cut in half and remove cores. Peel and cut into quarters or slices, if desired. Steam blanch pear halves 6 minutes or syrup blanch 10 minutes. Or, you can use an ascorbic acid, fruit juice, or sulfite dip.

Sun dry, 5 days
Dehydrator, 24 to 36 hours

PERSIMMON SLICES

Use firm fruit for long, soft varieties; use fully ripe fruit for round, drier varieties. See General Steps in Drying (pp. 218-219). Wash persimmons and drain. Use a stainless steel knife to peel and slice. Syrup blanch 10 minutes, if desired; pretreatment is not necessary.

Sun dry, 5 to 6 days
Dehydrator, 18 to 24 hours

PINEAPPLE SLICES

Use fully ripe, fresh pineapple. See General Steps in Drying (pp. 218-219). Wash and peel pineapple and remove the thorny eyes. Slice lengthwise and remove core. Cut crosswise into ½-inch slices. No pretreatment is necessary.

Dehydrator, 24 to 36 hours

PLUMS

Use prune plums for drying. See General Steps in Drying (pp. 218-219). Wash plums and drain. Leave whole or cut in half. No pretreatment is necessary.

Dehydrator, 24 to 36 hours

VEGETABLE RECIPES

ASPARAGUS

See General Steps in Drying (pp. 218-219). Wash asparagus thoroughly. Cut large tips in half. Steam blanch 4 to 5 minutes or water blanch 3½ to 4½ minutes. Drain well.

Sun dry, 8 to 10 hours
Oven dry, 3 to 4 hours
Dehydrator, 4 to 6 hours

SNAP BEANS

See General Steps in Drying (pp. 218-219). Wash beans thoroughly. Cut lengthwise or into short pieces. Steam blanch 2 to 2½ minutes or water blanch 2 minutes. Drain well. Freezing beans for 30 to 40 minutes after blanching will give them a better texture.

Sun dry, 8 hours
Oven dry, 3 to 6 hours
Dehydrator, 8 to 14 hours

BEET STICKS

See General Steps in Drying (pp. 218-219). Wash beets well. Trim tops but leave ½-inch stems and roots to prevent bleeding of color during cooking. Boil enough water to cover beets. Drop beets into boiling water and boil small beets 25 to 30 minutes; larger beets, 45 to 50 minutes. Cool beets, peel them, and cut into matchsticks, ⅛ inch thick.

Sun dry, 8 to 10 hours
Oven dry, 3 to 5 hours
Dehydrator, 10 to 12 hours

CABBAGE STRIPS

See General Steps in Drying (pp. 218-219). Remove outer leaves. Cut cabbage into quarters and remove core. Then, cut into strips ⅛ inch thick. Steam blanch 2½ to 3 minutes until wilted or water blanch 1½ to 2 minutes. Drain well.

Sun dry, 6 to 7 hours
Oven dry, 1 to 3 hours
Dehydrator, 10 to 12 hours

CARROT STRIPS OR SLICES

Use crisp, tender carrots. See General Steps in Drying (pp. 218-219). Wash carrots thoroughly and cut off roots and tops. Peel and cut into slices or strips ⅛ inch thick. Steam blanch 3 to 3½ minutes or water blanch 3½ minutes. Drain well.

Sun dry, 8 hours
Oven dry, 3½ to 5 hours
Dehydrator, 10 to 12 hours

SWEET CORN

Use tender, mature sweet corn. See General Steps in Drying (pp. 218-219). Remove husks and trim. Steam blanch 2 to 2½ minutes or water blanch 1½ minutes. Drain well. After blanching, cut kernels off cob.

Sun dry, 8 hours
Oven dry, 4 to 6 hours
Dehydrator, 12 to 15 hours

EGGPLANT SLICES

See General Steps in Drying (pp. 218-219). Wash and trim eggplant. Cut into ¼-inch slices. Steam blanch 3½ minutes or water blanch 4 to 5 minutes. Drain well.

Sun dry, 6 to 8 hours
Oven dry, 3½ to 5 hours
Dehydrator, 12 to 14 hours

CHOPPED GARLIC

See General Steps in Drying (pp. 218-219). Peel garlic and chop fine. No pretreatment is needed.

Dehydrator, 6 to 8 hours

GRATED HORSERADISH

See General Steps in Drying (pp. 218-219). Wash horseradish; remove rootlets and stubs. Peel or scrape root and grate. No pretreatment is needed.

Sun dry, 7 to 10 hours
Oven dry, 3 to 4 hours
Dehydrator, 4 to 10 hours

MUSHROOM SLICES

Do not use any wild mushrooms. See General Steps in Drying (pp. 218-219). Scrub thoroughly. Discard any tough, woody stalks. Cut tender stalks into short sections. Peel large mushrooms and slice. Do not peel small mushrooms. No pretreatment is needed.

Sun dry, 6 to 8 hours
Oven dry, 3 to 5 hours
Dehydrator, 8 to 10 hours

OKRA SLICES

See General Steps in Drying (pp. 218-219). Wash, trim, and cut okra crosswise into 1/8- to 1/4-inch slices. No pretreatment is needed.
Sun dry, 8 to 11 hours
Oven dry, 4 to 6 hours
Dehydrator, 8 to 10 hours

ONION SLICES

See General Steps in Drying (pp. 218-219). Wash onions and remove outer shells. Remove tops and root ends. Cut into 1/8- to 1/4-inch slices. No pretreatment is needed.
Sun dry, 8 to 11 hours
Oven dry, 3 to 6 hours
Dehydrator, 3 to 9 hours

PARSLEY

See General Steps in Drying (pp. 218-219). Wash parsley thoroughly. Separate clusters and discard long or tough stems. No pretreatment is needed.
Sun dry, 6 to 8 hours
Oven dry, 2 to 4 hours
Dehydrator, 1 to 2 hours

GARDEN PEAS

See General Steps in Drying (pp. 218-219). Wash hulls to remove loose dirt. Shell peas. Steam blanch 3 minutes or water blanch 2 minutes. Drain well.
Sun dry, 6 to 8 hours
Oven dry, 3 hours
Dehydrator, 8 to 10 hours

CHOPPED PEPPERS

See General Steps in Drying (pp. 218-219). Wash and remove stems. Cut in half and remove seeds and cores. Remove partitions. Cut peppers into 3/8-inch squares. No pretreatment is needed.
Sun dry, 6 to 8 hours
Oven dry, 2½ to 5 hours
Dehydrator, 8 to 12 hours

CHOPPED PIMIENTOS

See General Steps in Drying (pp. 218-219). Wash and remove stems. Cut in half and remove seeds and cores. Remove partitions. Cut pimientos into 3/8-inch squares. No pretreatment is needed.
Sun dry, 6 to 8 hours
Oven dry, 2½ to 5 hours
Dehydrator, 8 to 12 hours

POTATO STRIPS OR SLICES

See General Steps in Drying (pp. 218-219). Wash and peel potatoes. Cut them into shoestring strips ¼ inch thick or into ⅛-inch slices. Steam blanch 6 to 8 minutes or water blanch 5 to 6 minutes. Drain well.

Sun dry, 8 to 11 hours
Oven dry, 4 to 6 hours
Dehydrator, 8 to 12 hours

PUMPKIN STRIPS

See General Steps in Drying (pp. 218-219). Wash pumpkin. Cut or break it into pieces. Remove seeds and pulp from cavity. Cut rind into 1-inch strips and peel. Cut strips crosswise into pieces about ⅛ inch thick. Steam blanch 2½ to 3 minutes or water blanch 1 minute. Drain well.

Sun dry, 6 to 8 hours
Oven dry, 4 to 5 hours
Dehydrator, 10 to 16 hours

SUMMER SQUASH SLICES

See General Steps in Drying (pp. 218-219). Wash, trim, and cut squash into ¼-inch slices. Steam blanch 2½ to 3 minutes or water blanch 1½ minutes. Drain well.

Sun dry, 6 to 8 hours
Oven dry, 4 to 6 hours
Dehydrator, 10 to 12 hours

TOMATO WEDGES OR SLICES

See General Steps in Drying (pp. 218-219). Steam or dip in boiling water to loosen skins. Chill in cold water and remove skins. Slice tomatoes or cut them into wedges about ¾ inch wide. Cut small pear or plum tomatoes in half. Steam blanch 3 minutes or water blanch 1 minute. Drain well.

Sun dry, 8 to 10 hours
Oven dry, 6 to 8 hours
Dehydrator, 10 to 18 hours

MAKING FRUIT AND VEGETABLE LEATHERS

Fruit and vegetable leathers are a tasty, chewy, dried food. Most leathers are made from fruits, but a few vegetables are also appropriate for leathers. See Table 21 (below) for those best suited for making leathers.

Table 21. Suitability of Foods for Making Leathers

Food	Suitability
Apples	Excellent
Apricots	Excellent
Bananas	Fair to good
Blackberries	Excellent
Cherries	Excellent
Citrus Fruits	Only in combination
Citrus Peel	Only in combination
Coconuts	Only in combination
Crab Apples	Only in combination
Cranberries	Only in combination
Figs	Only in combination
Grapes	Fair to good
Melons	Not recommended
Nectarines	Not recommended
Peaches	Excellent
Pears	Excellent
Persimmons	Fair
Pineapples	Excellent
Plums	Good
Prune Plums	Excellent
Pumpkin	Good
Rhubarb	Fair
Strawberries	Excellent
Tomatoes	Good
Mixed Vegetables	Fair to good

Leathers are made by pouring pureed fruits or vegetables onto a flat surface and drying them. The name "leather" comes from the fact that when the pureed food is dry, it becomes shiny and has the texture of leather. When dried, the leather is rolled like a jelly roll to store.

Leftover fruit pulp from making jelly can be blended to make fruit leathers. For a diabetic, fruit leathers made without sugar are good snacks or desserts if they contain the amount of fruit allowed for the fruit exchange.

You can use fresh, frozen, or canned fruits and vegetables to make leathers.

GENERAL STEPS IN MAKING LEATHERS

1. Prepare fruit or vegetable into a puree (see Recipes, pp. 239-241).

2. Cover a baking sheet or tray with plastic wrap. A sheet with edges is better because the edges will keep the puree from spilling. Carefully smooth out any wrinkles in the wrap. Do not use wax paper or aluminum foil.

3. Pour about 2 cups of pureed fruit or vegetable mixture onto the baking sheet. Spread it evenly with a plastic spatula, close to but not touching the edges. It should be about ¼ inch thick.

4. Dry the leather: sun dry, 2 to 3 days; oven dry, 7 to 10 hours; dehydrator, 4 to 5 hours. (See How to Sun Dry, p. 226; How to Oven Dry, pp. 226-227; or How to Dehydrator Dry, p. 227).

5. Test for dryness by touching the center of the leather. No indention should be evident.

6. When leather is dry and still warm, roll it like a jelly roll.

7. Store in plastic bags. Leathers will keep up to 1 month at room temperature and up to 1 year in the freezer.

Spicing Up Leathers

The following spices, flavorings, garnishes, and fillings are appropriate for some fruit and vegetable leathers. Try the ones you think might be good.

Spices. Sprinkle leathers with a ground spice, such as allspice, cinnamon, cloves, coriander, ginger, mace, mint, nutmeg, pumpkin pie spice. Start with ⅛ teaspoon for each 2 cups of puree. Add more, according to your taste.

Flavorings. Try the ones that you think go best with the fruit or vegetable you are using: vanilla extract, almond extract, orange extract, lemon juice, lime juice, orange juice, grated lemon peel, grated lime peel, grated orange peel. Try ⅛ to ¼ teaspoon for each 2 cups of puree.

Garnishes. Use one or more, according to your taste: shredded coconut, chopped dates, chopped raisins, chopped nuts, granola, miniature marshmallows, poppy seeds, sesame seeds, sunflower seeds. Sprinkle over leather just before rolling.

Fillings. Spread your favorites over the leather just before rolling: melted chocolate, softened cream cheese, cheese spreads, jam, preserves, marmalade, marshmallow creme, peanut butter. Store in refrigerator.

PUREE RECIPES FOR LEATHERS

FRESH FRUIT PUREE
Makes enough for 1 cookie sheet.

See General Steps in Making Leathers (p. 238). Use ripe or slightly overripe fruit. Wash in cool water. Peel, if needed, and remove seeds and stems. Cut fruit into chunks. Put chunks into a food blender and puree. Prepare enough to make 2 cups puree. For light-colored fruits, add 2 teaspoons lemon juice or ⅛ teaspoon ascorbic acid (375 mg) and mix well.

Note: If you do not have a blender, combine fruit and its juice and cook over low heat until soft. Then, mash with a pastry blender or potato masher.

To Vary: Add ¼ to ½ cup corn syrup, honey, or sugar. Syrup and honey are better for longer storage because they prevent crystal formation. Saccharin-based sweeteners can be used; they reduce tartness without adding calories. Aspartame-based sweeteners may lose sweetness during drying.

CANNED OR FROZEN FRUIT PUREE
Makes enough for 1 cookie sheet.

See General Steps in Making Leathers (p. 238). Drain canned fruit and save the juice. Put 2 cups canned or frozen fruit pieces into a food blender and puree. Or, mash with a pastry blender or potato masher until smooth. If puree is too thick, add some juice. For yellow or light-colored fruits, add 2 teaspoons lemon juice or ⅛ teaspoon ascorbic acid (375 mg) and mix well.

Note: Canned applesauce can be dried alone or added to any fresh fruit puree. It decreases tartness and makes the leather smoother and more pliable.

SPICED APPLE PUREE
Makes enough for 2 cookie sheets.

4 medium apples
¼ cup light corn syrup
⅛ teaspoon ground cinnamon
⅛ teaspoon ground cloves
1 tablespoon lemon juice

Use fully ripe or overripe apples. See General Steps in Making Leathers (p. 238). Wash, peel, and core apples and cut into small chunks. Prepare enough to make 4 cups of apple chunks. Put 2 cups apples into a food blender and add syrup, cinnamon, cloves, and lemon juice. Cover and puree them until mixture looks like applesauce. Add the remaining 2 cups of apples and puree them.

FRESH PEACH PUREE
Makes enough for 2 cookie sheets.

4 medium peaches
1 tablespoon lemon juice
2 tablespoons sugar

Use fully ripe or overripe peaches. See General Steps in Making Leathers (p. 238). Wash and peel peaches. Cut in half and remove seeds. Cut peaches into small chunks. Prepare enough to make 4 cups peach chunks. Put 2 cups peaches into a food blender and add lemon juice and sugar. Cover and puree until all chunks are gone. Add the remaining 2 cups of peaches and puree until mixture is smooth.

PERSIMMON PUREE
Makes enough for 1 cookie sheet.

Use fully ripe persimmons. See General Steps in Making Leathers (p. 238). Wash and peel persimmons. Prepare enough to make 4 cups persimmons. In a thick-bottom saucepan, combine persimmons with 1 cup water and heat to a boil, stirring to prevent sticking. Boil until it makes a thick puree. Spread a thin layer of puree on plastic wrap and remove seeds. Dry it. Then, add another thin layer and repeat until leather is thick enough to handle easily.

PUMPKIN PUREE
Makes enough for 1 cookie sheet.

2 cups canned pumpkin
½ cup honey
¼ teaspoon cinnamon
⅛ teaspoon nutmeg
⅛ teaspoon powdered cloves

See General Steps in Making Leathers (p. 238). Combine all ingredients and mix well. Or, you can use fresh pumpkin. Prepare and cook enough to make 2 cups of puree and add remaining ingredients.

FROZEN STRAWBERRY PUREE
Makes enough for 2 cookie sheets.

1 10-ounce package frozen sweetened strawberries
1 tablespoon lemon juice

See General Steps in Making Leathers (p. 238). Put frozen strawberries into a food blender and add lemon juice. Puree until the mixture is smooth. Or, place frozen strawberries in a large bowl, add lemon juice, and mash with a pastry blender or potato masher until mixture is smooth.

TOMATO PUREE
Makes enough for 1 cookie sheet.

4 medium tomatoes　　　　　　　　Salt to taste

See General Steps in Making Leathers (p. 238). Wash tomatoes and remove cores. Cut tomatoes into quarters. Place them in a saucepan, cover, and cook over low heat for 15 to 20 minutes. Pour them into a food blender and puree or press them through a sieve. Pour pureed tomatoes into an electric skillet or shallow pan. Add salt to taste and cook over low heat until tomatoes are thick.

MIXED VEGETABLE PUREE
Makes enough for 1 cookie sheet.

3 medium tomatoes　　　　　　　　¼ cup chopped celery
1 small onion　　　　　　　　　　　¼ teaspoon salt, or to taste

Use ripe tomatoes. See General Steps in Making Leathers (p. 238). Wash and core tomatoes and cut them into small pieces. Prepare enough to make 2 cups. Wash, peel, and chop onion. Put tomatoes, onion, and celery in a saucepan. Cover and cook over low heat for 15 to 20 minutes. Pour mixture into a food blender and puree or press through a sieve. Pour pureed mixture into an electric skillet or shallow pan. Add salt and cook over low heat until mixture is thick.

DRYING HERBS

Drying

Air Drying (p. 221) is the best method of drying for most herbs. Drying in the sun, an oven, or a dehydrator can reduce flavor and color.

Testing for Dryness

Place dried herbs in tightly sealed jars in a warm place for about 1 week. Then, check the jars for moisture. If moisture accumulates on the inside of the glass or under the lid, remove the herbs and dry them some more. Otherwise, mold may develop.

Storing

Pack herbs in dark-colored glass jars or in tins cans or boxes. Do not use cardboard or paper containers because paper products absorb oils and leave the herbs tasteless. Store them in a cool place, out of strong light.

HERB RECIPES

HERB LEAVES

Cut plants for leaves just before blossoming, when the plant's stock of essential oils is at its highest. Cut them early on hot, dry days, just as soon as the dew is off the plant. Cut off the tip growth—about 6 inches of stems below the flower buds. Do not wash leaves unless they are dusty or have been thickly mulched. Then, rinse them in cool water and shake gently to remove excess water. Be careful not to bruise the leaves. Discard all that are bruised, soiled, or imperfect.

Bunch Dried. *For basil, chervil, marjoram, parsley, rosemary, sage, and savory.* Tie the herbs in small bunches with their leaves down so oils will flow from the stems into the leaves. Hang them in the sun just long enough for any water to evaporate from them. Then, hang them in a warm, dry, well-ventilated room or porch away from bright light. To prevent dust contamination, tie a brown paper bag over each bunch. Cut or punch holes in the bag to allow air circulation. Herbs are best if dried in 3 or 4 days. If they are not entirely dry in 2 weeks, place them in a closed oven until thoroughly dry. If using an electric oven, turn on the oven light; it will provide enough heat to dry the leaves. The pilot light of a gas oven is warm enough.

Leaf Dried. *For bay, chervil, lemon verbena, mint, rosemary, and thyme.* These may be dried by removing the best leaves from the stems and spreading them in single layers on paper towels on a tray. The leaves should not touch each other. Up to five layers of paper towels with leaves can be stacked on each other. Dry them in a cool oven overnight. The oven light of an electric oven or the pilot light of a gas oven provides enough heat to dry individual leaves. They dry flat and keep their color well.

When leaves are crispy dry and crumple easily between your fingers, they are ready to be packaged and stored.

HERB SEEDS

For anise, caraway, coriander, cumin, dill, and fennel. Cut plants when seed pods or heads have changed color but before they begin to shatter. There are two ways to dry seeds.

Pod Dried. Spread the pods in single layers on screen or net trays in a warm, dry room or porch. When pods are thoroughly dry, rub them between the palms of your hands to remove the seeds.

Bag Dried. Hang each whole plant upside down inside a clean paper bag. Hang the bags in a warm, dry room or porch. Cut or punch holes in the sides of the bag to allow air circulation. The bag will catch the seeds as they dry and fall from the pods.

HERB ROOTS

For angelica, burdock, comfrey, ginger, ginseng, and sassafras. Dig for roots during plant's dormant stage, usually during fall and winter. That is when enough food is stored in the plant to keep it from dying. Cut only a few tender roots from each plant; repack soil around remaining roots. Scrub roots with a vegetable brush to remove dirt. Leave thin roots whole; slice thick roots lengthwise for quicker drying. Oven drying (see How to Oven Dry, pp. 226-227) is best for roots. Drying should take only a few hours.

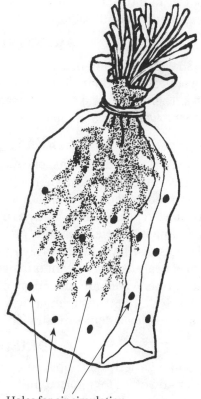

Holes for air circulation

Figure 18. Drying Herbs in a Paper Bag

DRYING PEANUTS, POPCORN, AND SEEDS

Drying and roasting are two different processes. Drying is not a form of cooking; it is a form of preserving. Roasting is a cooking process. The recipes for peanuts and seeds will include drying and roasting. Read General Steps in Drying (pp. 218-219) before beginning to dry these products.

DRIED PEANUTS

Unshelled peanuts can be oven dried (see How to Oven Dry, pp. 226-227) in a single layer on a baking sheet at 130°F. Or, they may be dried in a dehydrator (see manufacturer's instructions). Peanuts are dry when their shells have hardened to a brittle state but the nutmeat is tender and not shriveled.

BLANCHED PEANUTS

Shell peanuts and place them in a large bowl or pot. Boil enough water to cover peanuts. Add boiling water to peanuts and let them sit for 2 to 4 minutes. Then, drain and add cold water. Remove skins and let peanuts dry on paper towels before using or storing.

ROASTED PEANUTS

To Roast Dried Peanuts, place a single layer of unshelled dried peanuts in a shallow pan and roast them in the oven at 300°F for 30 to 40 minutes. Or, shell dried peanuts and roast them in the oven at 300°F for 20 to 25 minutes. Stir shelled peanuts frequently to prevent scorching.

To Roast Undried Peanuts, preheat oven to 300°F. Shell raw peanuts and place a single layer in a baking pan. If desired, add ¼ cup margarine for each cup of peanuts. Place peanuts in oven and roast for 25 to 35 minutes, if using margarine. When nuts are brown, pour them onto absorbent paper. Add salt, if desired. If not using margarine, roast peanuts for 35 to 45 minutes or until nuts are brown. Remove skins, if desired.

DRIED POPCORN

Some varieties of popcorn can be dried at home. Contact your local Extension office for information. Leave the ears of corn on the stalks until the kernels are well dried. Then, remove ears and allow kernels to air dry on the cobs. When kernels are dry, they will appear shriveled. Remove them from the ears and package them for storage.

DRIED PUMPKIN SEEDS

Carefully wash pumpkin seeds to remove any clinging fibrous pumpkin tissue. Seeds can be sun dried (see How to Sun Dry, p. 226); they can be dried in a dehydrator at 115° to 120°F for 1 to 2 hours; or they can be oven dried (see How to Oven Dry, pp. 226-227) on the oven's warm setting for 3 to 4 hours. Stir seeds in the oven frequently to avoid scorching.

ROASTED PUMPKIN SEEDS

Preheat oven to 250°F. Mix dried seeds with just enough oil to lightly coat seeds. Add salt, if desired. Place in a flat pan in the oven for 10 to 15 minutes. Stir occasionally.

DRIED SUNFLOWER SEEDS

Sunflower seeds are dried best while on the flower. The flower may have to be wrapped with cheesecloth to prevent birds from eating the seeds. Or, seeds may be removed and sun dried or dried in a dehydrator at 100°F. Sunflower seeds develop an off-flavor when dried at higher temperatures.

ROASTED OR FRIED SUNFLOWER SEEDS

Dried seeds may be roasted in a shallow pan at 300°F for 10 to 15 minutes. Or, raw, mature seeds may be soaked in a salt-water solution and roasted or deep-fat fried.

To Soak Seeds: Cover unshelled seeds with 2 quarts water and ¼ to ½ cup salt. Bring to a boil and simmer 2 hours. Drain seeds and dry on paper towels.

To Roast Seeds: Measure 2 cups treated sunflower seeds into a shallow pan and spread evenly. Place pan in a 300°F oven and bake for 30 to 40 minutes. Stir occasionally. When seeds are golden brown, remove from oven and add 2 teaspoons melted butter, margarine, or vegetable oil. Stir to coat. Pour out onto paper towels and salt to taste.

To Fry Seeds: Heat ¼ cup vegetable oil for each cup of seeds in a deep-fat fryer or electric skillet set at 360°F. Measure seeds into hot oil and cook until golden brown. Skim off seeds and place on paper towels to drain. Add salt to taste.

Store seeds in a clean container with a tight-fitting lid.

FOOD PRESERVATION TIMETABLES

Table 22. Canning Timetable for High-Acid Fruits
Using a boiling water-bath canner (212°F).

Food*	Type Pack	Minutes	
		Pints	Quarts
Apple Juice	Hot	5	5
Apples	Hot	20	20
Applesauce	Hot	15	20
Apricots	Hot	20	25
	Raw	25	30
Berries	Hot	15	15
	Raw	15	20
Berry Juice	Hot	5	5
Cherries	Hot	15	20
	Raw	25	25
Cranberries	Hot	15	15
Figs	Hot	45	50
Grape Juice	Hot	5	5
Grapes	Hot	10	10
	Raw	15	20
Grapefruit	Raw	10	10
Nectarines	Hot	20	25
	Raw	25	30
Peaches	Hot	20	25
	Raw	25	30
Pears	Hot	20	25
Persimmon Sauce			
Japanese	Hot	5	10
Native	Hot	15	20
Pineapple	Hot	15	20
Plums	Hot/Raw	20	25
Rhubarb	Hot	15	15
Tomatoes	Hot/Raw	35	40
Tomato Juice	Hot	35	40
Tomato Puree	Hot	35**	40

*Check recipes for headspace (pp. 42-70).
**For pints and half-pints.

Table 23. Canning Timetable for Low-Acid Vegetables
Using a pressure canner (240°F) at 10 pounds pressure.

Food*	Type Pack	Minutes Pints	Minutes Quarts
Asparagus	Hot/Raw	30	40
Lima Beans	Hot/Raw	40	50
Snap Beans	Hot/Raw	20	25
Soy Beans	Hot	55	65
Beets	Hot	30	35
Carrots	Hot/Raw	25	30
Corn, cream-style	Hot	85	**
	Raw	95	**
Corn, whole-kernel	Hot/Raw	55	85
Greens, all kinds	Hot	70	90
Okra	Hot	25	40
Garden Peas	Hot/Raw	40	40
Southern Peas	Hot/Raw	40	50
Pimientos	Hot	35	(½ pint only)
Irish Potatoes	Hot	35	40
Pumpkin, cubed	Hot	55	90
Summer Squash	Hot	30	40
	Raw	25	30
Winter Squash	Hot	55	90
Sweet Potatoes	Hot, dry	65	95
	Hot, wet	55	90
Vegetable Soup	Hot	Time needed for vegetable requiring longest time	

*Check recipes for headspace (pp. 72-86).

Table 24. Canning Timetable for Meats, Poultry, and Fish
Using a pressure canner (240°F) at 10 pounds pressure.

Food*	Type Pack	Minutes Pints	Minutes Quarts
Beef, Veal, Pork, Lamb (no bone)	Hot/Raw	75	90
Ground, Chopped Beef**	Hot	75	90
Poultry, Rabbit (with bone)	Hot/Raw	65	75
Poultry, Rabbit (without bone)	Hot/Raw	75	90
Fish (salt and fresh)	Raw	100	***

*Check recipes for headspace (pp. 88-92).
**Raw pack is not recommended for ground beef because it is hard to remove from jars.
***Quarts not recommended.

Table 25. Blanching Timetable for Freezing Vegetables

Vegetable	Blanching* Method	Time** (minutes)
Asparagus		
Small spears	Water	2
Medium spears	Water	3
Large spears	Water	4
Lima Beans		
Small beans	Water	2
Medium beans	Water	3
Large beans	Water	4
Snap Beans	Water	3
Beets	Boil until tender	
Broccoli	Steam	5
	Water	3
Cabbage or Chinese Cabbage	Water	1½
Carrots, sliced	Water	2
Carrots, whole	Water	5
Cauliflower	Water	3
Corn, cream-style	Water	4
Corn, on cob		
Small ears	Water	7
Medium ears	Water	9
Large ears	Water	11
Corn, whole-kernel	Water	4
Eggplant, sliced	Water	4
Greens		
Collards	Water	3
All others	Water	2
Okra		
Small pods	Water	3
Large pods	Water	4
Onions, cut	Not necessary	
Onions, whole	Water	3 - 7
Garden Peas	Water	1½
Southern Peas	Water	2
Hot Pepper	Not necessary	
Sweet Pepper	Not necessary	
Pimiento	Water	3 - 5
Irish Potatoes		
new, ¾ inch	Water	4
1 inch	Water	6
1½ inch	Water	7
Pumpkin	Boil until tender	
Summer Squash	Water	3
Winter Squash	Boil until tender	
Sweet Potatoes, sliced, cured	Water	5
Sweet Potatoes, for puree	Boil until tender	
Tomatoes	Cook 10 - 20 minutes	

*Follow instructions for blanching (pp. 194-195). Generally, for water blanching, use 1 gallon water per pint or pound for most vegetables; use 2 gallons water for leafy vegetables. For steam blanching, place vegetables one layer deep on a rack 3 inches above 1 to 2 inches of rapidly boiling water.
**Count blanching time from the moment water returns to boiling point after adding vegetables.

Table 26. Blanching and Drying Timetable for Fruits*

Fruit	Blanching Time**		Drying Time***	
	Method	Minutes	Method+	Hours
Apples	Steam depending on texture	5	Dehydrator Sun	6 - 12 36 - 48
Apricots	Steam Water	3 - 4 4 - 5	Dehydrator Sun	24 - 36++ 24 - 36
Figs	Not necessary		Dehydrator Sun	12 - 20 48 - 60
Grapes, seedless	Not necessary		Dehydrator Sun	12 - 20 36 - 60
Nectarines	Steam Water	8 8	Dehydrator Sun	36 - 48++ 36 - 60
Peaches	Steam Water	8 8	Dehydrator Sun	36 - 48++ 36 - 60
Pears	Steam	6	Dehydrator Sun	24 - 36 60
Persimmons	Not necessary		Dehydrator Sun	18 - 24 60 - 72

*Check recipes for drying fruits (pp. 230-232).
**Fruits may be sulfured instead of blanched. Contact your local Extension office for information.
***To test fruits for dryness, see Testing for Dryness (pp. 227-228).
+Range ovens may be used, but the length of drying time for fruits makes it impractical for any but apples.
++Drying time can be shortened by cutting fruit into slices.

Table 27. Blanching and Drying Timetable for Vegetables*

Vegetable	Blanching Time Method	Minutes	Drying Time** Method	Hours
Asparagus	Steam	4 - 5	Dehydrator	6 - 10
	Water	3½ - 4½	Oven	3 - 4
			Sun	8 - 10
Lima Beans	Steam	4	Dehydrator	8 - 14
	Water	4	Oven	3 - 6
			Sun	8
Snap Beans	Steam	2 - 2½	Dehydrator	8 - 14
	Water	2	Oven	3 - 6
			Sun	8
Beets	Cook before drying		Dehydrator	10 - 12
			Oven	3 - 5
			Sun	8 - 10
Broccoli	Steam	3 - 3½	Dehydrator	12 - 15
	Water+	2	Oven	3 - 4½
			Sun	8 - 10
Brussels Sprouts	Steam	6 - 7	Dehydrator	12 - 18
	Water	4½ - 5½	Oven	4 - 5
			Sun	9 - 11
Cabbage	Steam until wilted	2½ - 3	Dehydrator	10 - 12
			Oven	1 - 3
	Water	1½ - 2	Sun	6 - 7
Carrots	Steam	3 - 3½	Dehydrator	10 - 12
	Water	3½	Oven	3½ - 5
			Sun	8
Cauliflower	Steam	4 - 5	Dehydrator	12 - 15
	Water+	3 - 4	Oven	4 - 6
			Sun	8 - 11
Celery	Steam	2	Dehydrator	10 - 16
	Water	2	Oven	3 - 4
			Sun	8
Corn, cut	Blanch before cutting from cob		Dehydrator	6 - 10
			Oven	2 - 3
			Sun	6
Corn, on cob and whole grain	Steam until milk is set	2 - 2½	Dehydrator	12 - 15
			Oven	4 - 6
	Water	1½	Sun	8
Eggplant	Steam	3½	Dehydrator	12 - 14
	Water	3	Oven	3½ - 5
			Sun	6 - 8
Greens, all kinds	Steam until thoroughly wilted	2 - 2½	Dehydrator	8 - 10
			Oven	2½ - 3½
	Water	1½	Sun	6 - 8
Horseradish	Not necessary		Dehydrator	4 - 10
			Oven	3 - 4
			Sun	7 - 10
Mushrooms (WARNING, see ++)	Not necessary		Dehydrator	8 - 10
			Oven	3 - 5
			Sun	6 - 8

Table 27. Blanching and Drying Timetable for Vegetables (cont.)

Vegetable	Blanching Time		Drying Time**	
	Method	Minutes	Method	Hours
Okra	Not necessary		Dehydrator	8 - 10
			Oven	4 - 6
			Sun	8 - 11
Onions	Not necessary		Dehydrator	3 - 9
			Oven	3 - 6
			Sun	8 - 11
Parsley	Not necessary		Dehydrator	1 - 2
			Oven	2 - 4
			Sun	6 - 8
Peas, all kinds	Steam	3	Dehydrator	8 - 10
	Water	2	Oven	3
			Sun	6 - 8
Peppers and Pimientos	Not necessary		Dehydrator	8 - 12
			Oven	2½ - 5
			Sun	6 - 8
Potatoes	Steam	6 - 8	Dehydrator	8 - 12
	Water	5 - 6	Oven	4 - 6
			Sun	8 - 11
Summer Squash	Steam	2½ - 3	Dehydrator	10 - 12
	Water	1½	Oven	4 - 6
			Sun	6 - 8
Winter Squash	Steam	2½ - 3	Dehydrator	10 - 16
	Water	1	Oven	4 - 5
			Sun	6 - 8
Tomatoes, for stewing	Steam	3	Dehydrator	10 - 18
	Water	1	Oven	6 - 8
			Sun	8 - 10

*Check recipes for drying vegetables (pp. 233-236).
**To test vegetables for dryness, see Testing for Dryness (pp. 227-228).
+Preferred method.
++WARNING: The toxins of poisonous varieties of mushrooms are not destroyed by drying or cooking. Only an expert can tell the difference between poisonous and edible varieties. Do not use wild mushrooms.

REFERENCES

Complete Guide To Home Canning, Extension Service, U.S.D.A., 1988.

So Easy To Preserve, Cooperative Extension Service, The University of Georgia, 1993.

Using Florida Citrus Fruits, Florida Cooperative Extension Service, University of Florida.

Index

A

Acid Crystals, Tartaric, 49, 127, 130
Acid Foods
  Canning, 17
  Glossary, 6
  Table 2, 11
Acid Test, Jelly, 128
Acidification
  Glossary, 6
  Tomatoes, 59
Air Drying, 221
Alcohol Test, Jelly, 128
Altitude
  Canning, 18
  Glossary, 6
  Table 3, 11
Alum, 96
Amber Marmalade, 148
Angelica, drying, 243
Anise, drying, 243
Apples
  Apple Butter, 156
  Apple Jelly, 132
  Apple Juice, canning, 42
  Applesauce, canning, 43
  Canning, 43
  Drying, 230
  Freezing, 186
  Pie Filling, canning, 54
  Refrigerated Apple Spread, 162
  Spiced Apple Puree, 239
  Spiced Apple Rings, 119
Applesauce, canning, 43
Apricots
  Canning, 44
  Drying, 230
Ascorbic Acid
  Drying, 225
  Freezing, 183-184
  Glossary, 6
  Solutions, 39
Asparagus
  Canning, 72
  Drying, 233
  Freezing, 197

B

Bacteria
  Clostridium botulinum,
    Canning, 17-18
    Glossary, 7
    Making Pickles, 94, 95
  E. coli, 218
  Glossary, 6
Banana Peppers (see Peppers, Hot)
Bananas
  Drying, 230
  Freezing, 187
Basil, drying, 242
Bay Leaves, drying, 243
Beans
  Dried
    Baked Beans, canning, 73
    Pickled Three-Bean Salad, 112
    Plain, canning, 73
    With Tomato Sauce, canning, 74
  Garbanzo
    Pickled Three-Bean Salad, 112
  Lima (Butter, Pinto, Soy)
    Canning, 74
    Freezing, 197
    Mixed Vegetables, canning, 84
    Succotash, canning, 83
  Snap (Green, Italian, Wax)
    Canning, 75
    Drying, 233
    Freezing, 197
    Mixed Vegetables, canning, 84
    Pickled Dilled Beans, 112
    Pickled Three-Bean Salad, 112
Beef
  Canning, 88-89
  Freezing, 203-205
Beets
  Canning, 75
  Drying, 233
  Freezing, 197
  Pickled Beets, 113
Bell Peppers (see Peppers, Sweet)
Berries (also see specific berries)
  Canning, 44
  Drying, 230
  Freezing, 187-188
  Juice, canning, 45
  Making Jelly, 133
Berry Syrup, 161
Blackberries
  Blackberry Jam, 139
  Blackberry Jelly, 133
Blanchers
  Drying, 221
  Freezing, 175
  Glossary, 6
Blanching
  Steam
    Drying, 224
    Freezing, 195

Syrup, 225
Water
  Drying, 224
  Freezing, 194
Blender Ketchup, 68
Blueberries
  Blueberry Jelly, 133
  Blueberry Spice Jam, 140
  Freezing, 188
  Pie Filling, canning, 55
Boiling Water-Bath Canner
  Canning, 26-28
  Glossary, 6
Boiling Water-Bath Processing
  Canning, 32
  Making Pickles, 96
Botulism
  Canning, 17-18
  Glossary, 7
Bread and Butter Pickles, 103
Breads, freezing, 212
Broccoli, freezing, 198
Broth
  Meat, canning, 89
  Poultry, canning, 90
Builder's Lime, 96
  Bread and Butter Pickles, 103
  Quick Sweet Pickles, 105
Burdock, drying, 243
Butcher Wrap, 178
Butter, freezing, 211
Butters, Fruit, 155-160
  Apple Butter, 156
  Crab Apple Butter, 156
  Grape Butter, 157
  Peach Butter, 157
  Peach Pineapple Butter, 158
  Pear Butter, 158
  Persimmon Butter, 159
  Plum Butter, 159
  Quince Butter, 160

C

Cabbage
  Dixie Relish, 110
  Drying, 233
  Freezing, 198
  Garden Relish, 110
  Piccalilli, 108
  Sauerkraut, 100
Canners
  Pressure, 28-30
  Using the Pressure, 32-34
  Using the Water-Bath, 32
  Water-Bath, 28

Canning
  Fairs and Exhibits, 163-167
  Fruits, 39-54
  Glossary, 7
  Meat, 87-89
  Nuts, 93
  Pie Fillings, 54-58
  Poultry, 89-90
  Seafood, 90-92
  Tomatoes, 59-70
  Vegetables, 71-86
Canning Salt
  Canning, 71
  Glossary, 7
  Making Pickles, 95
Caraway, drying, 243
Carrots
  Canning, 76
  Carrot Orange Marmalade, 149
  Drying, 234
  Freezing, 198
  Mixed Vegetables, canning, 84
Cauliflower
  Freezing, 198
  Garden Relish, 110
  Pickled Cauliflower, 113
Cheese, freezing, 211
Cherries
  Canning, 46
  Drying, 231
  Pie Filling, canning, 56
  Sour, Whole, freezing, 188
  Sweet, Whole, freezing, 188
Chervil, drying, 242, 243
Chestnuts, Shelled, canning, 93
Chickens (also see Poultry)
  Canning, 89
  Freezing, 205
Chili Peppers (see Peppers, Hot)
Chili Salsa, canning, 70
Chinese Cabbage, freezing, 198
Citric Acid
  Canning, 17-18, 59
  Glossary, 7
Citrus Fruit Marmalade, 149
Citrus Peel, 231
Clostridium botulinum
  Canning, 17-18
  Glossary, 7
  Making Pickles, 94
Clams
  In Shell, freezing, 207
  Shucked
    Canning, 90
    Freezing, 207

Cold Pack, 7
Collards
  Canning, 77
  Freezing, 199
Comfrey, drying, 243
Conditioning Fruits, 228
Conserves, 152-155
  Cranberry Conserve, 153
  Damson Plum Conserve, 153
  Fig Conserve, 154
  Grape Conserve, 154
  Pear Conserve, 155
  Pear Muscadine Conserve, 155
Containers (see Packaging Materials)
Cooked Foods, freezing, 212-216
  Breads, 212
  Desserts, 213
  Meat and Poultry, 214
  Salads, 215
  Sandwiches, 215
  Soups, 215
  Vegetables, 216
Cooking Test, Jelly, 127
Coriander, drying, 243
Corn
  Corn on the Cob, freezing, 199
  Corn Relish, 109
  Cream-Style Corn
    Canning, 76
    Freezing, 198
  Mixed Vegetables, canning, 84
  Popcorn, drying, 244
  Succotash, canning, 83
  Whole-Kernel Corn
    Canning, 77
    Drying, 234
    Freezing, 199
Country Western Ketchup, 67
Crab Apples
  Crab Apple Butter, 156
  Crab Apple Jelly, 134
  Spiced Crab Apples, 119
Crabs
  In Shell, freezing, 207
  Shelled
    Canning, 91
    Freezing, 207
Cranberries
  Canning, 46
  Cranberry Conserve, 153
Cream, Whipping, freezing, 211
Crock, Stoneware, 97-98
Crushed Tomatoes, 62
Cucumbers
  Bread and Butter Pickles, 103
  Dill Pickles, 101
  14-Day Sweet Pickles, 104
  Pickle Relish, 107
  Pickled Mixed Vegetables, 118
  Quick Fresh-Pack Dill Pickles, 102
  Quick Sweet Pickles, 105
  Refrigerator Dills, 102
  Reduced-Sodium Sliced Dill Pickles, 103
  Reduced-Sodium Sliced Sweet Pickles, 107
  Sweet Gherkin Pickles, 106
Cumin, drying, 243

D

Dairy Foods, freezing, 211
Damson Plum Conserve, 153
Dark and Spicy Eggs, 123
Dehydrator Drying, 220
Dehydrators, 227
Desserts, Cooked, freezing, 213
Dill, drying, 243
Dill Pickles, 101
Dilled Eggs, 123
Dips for Drying
  Ascorbic Acid, 225
  Fruit Juice, 225
  Honey, 225
  Saline Solution, 225
  Sulfite, 225
Dixie Relish, 110
Drugstore Wrap, 176-177
Dry Pack
  Freezing Fruits, 184
  Freezing Vegetables, 196
Dryers, 219-221, 226-227
Drying, 217-245
  Conditioning Fruits, 228
  Fruits, 230-232
  General Steps, 218-219
  Herbs, 242-243
  Making Leathers, 237-241
  Materials and Equipment, 219-222
  Methods and Procedures, 222-230
  Nuts, 244
  Packaging, 228-229
  Pasteurizing, 228
  Popcorn, 244
  Reconstituting
    Fruits, 229
    Vegetables, 229
  Seeds, 245
  Storing, 229
  Testing for Dryness, 227-228
  Vegetables, 233-236

E

E. coli Bacteria, 218
Eggplant
  Drying, 234
  Freezing, 199
Eggs
  Dark and Spicy Eggs, 123
  Dilled Eggs, 123
  Freezing, 209
  Pickled Eggs, 122
  Red Beet Eggs, 123
  Rosy Pickled Eggs, 123
  Spicy Eggs, 123
  Sweet and Sour Eggs, 123
Elderberries, freezing, 188
Enzymes
  Freezing, 168, 195
  Glossary, 7
Equipment
  Canning, 21-30
  Drying, 219-222
  Freezing, 170-175
  Making Jelly, 130
  Making Pickles, 98
Equivalents
  Table 4, 12
  Table 5, 13
  Table 6, 14
  Table 7, 14
  Table 8, 15
Exhausting
  Canning, 29, 33
  Glossary, 7
Extracting Juice, 126-127

F

Fennel, drying, 243
Fermentation
  Glossary, 8
  Making Pickles, 97-98
    Dill Pickles, 101
    14-Day Sweet Pickles, 104
    Sauerkraut, 100
    Sweet Gherkin Pickles, 106
Figs
  Canning, 47
  Fig Conserve, 154
  Drying, 231
  Fig Jam, 140
  Fig Preserves, 143
  Freezing, 188
  Strawberry Fig Preserves, 147
Firming Agents, 96

Fish (also see Seafood)
  Canning
    Fillets, 91
    Smoked, 92
    Freezing, 206
Fish Roe, freezing, 207
Food Quality
  Canning, 18-19
  Drying, 218
  Freezing, 168-169
Food Safety
  Canning, 17-18
  Drying, 218
  Freezing, 168-169
14-Day Sweet Pickles, 104
Freezer Burn, 169
Freezers, 173-175
  Care, 173-175
  Defrosting, 174
Freezing, 168-216
  Cooked Foods, 212-216
  Dairy Foods, 211,
  Eggs, 209
  Equipment, 170-175
  Fruits, 183-193
  Keeping Records, 179
  Meats, 203-205
  Nuts, 210
  Poultry, 205-206
  Problems, 181-182
  Procedures, 175-181
  Refreezing Food, 179-181
  Seafood, 206-208
  Vegetables, 194-203
Fruit Cocktail, canning, 47
Fruit Juice Dip, 225
Fruit Leathers, 237-241
Fruit Purees
  Canning, 48
  For Leathers, 239-240
Fruit Syrup, 161
Fruits
  Canning, 39-58
  Drying, 230-232
  Freezing, 183-193
  For Leathers, 239-240
  Making Jellies, 124-162
  Making Pickles, 119-121

G

Game, Wild
  Canning, 88
  Freezing, 205

Game Birds
  Canning, 89-90
  Freezing, 206
Garden Relish, 110
Garlic, drying, 234
General Steps
  Canning, 19-20
  Drying, 218-219
  Freezing, 169-170
  Making Jellied Spreads, 138-139
  Making Jelly, 126-129
  Making Leathers, 238
Ginger, drying, 243
Gingered Pear Preserves, 145
Ginseng, drying, 243
Glazes, Seafood, freezing
  Ice Glaze, 206
  Lemon Gelatin Glaze, 206
Glossary, 6-9
Grapefruit
  Amber Marmalade, 148
  Canning, 48
  Citrus Fruit Marmalade, 149
  Drying peel, 231
  Freezing, 189
Grapes
  Canning, 49
  Drying, 231
  Freezing, 189
  Grape Butter, 157
  Grape Conserve, 154
  Grape Jam, 140
  Grape Jelly, 134
  Grape Juice, canning, 48
  Grape Marmalade, 149
  Grape Plum Jelly, 135
  Refrigerated Grape Spread, 162
Green Tomatoes
  Freezing, 203
  Green Tomato Marmalade, 150
  Pickled Green Tomato Relish, 109
  Pickled Sweet Green Tomatoes, 117
  Green Tomato Pie Filling, canning, 57
Greens
  Canning, 77
  Freezing, 199

H

Headspace
  Canning, 31-32
  Freezing, 176
  Glossary, 8
  Table 14, 179
Heat Processing, 8

Herbs
  Drying
    Leaves, 242-243
    Parsley, 235
    Roots, 243
    Seeds, 243
  Freezing, 199
Honey Dip, 225
Horseradish
  Drying, 234
  Pickled Horseradish Sauce, 111
Hot Pack
  Canning, 30
  Glossary, 8
How to Win at Canning, 164-167
Huckleberries, freezing, 188
Hungarian Peppers (see Peppers, Hot)

I

Ice Formation, 169
Ice Glaze, 206
Italian Beans (see Beans, Snap)

J

Jalapeno Pepper Jelly, 137
Jalapeno Peppers (see Peppers, Hot)
Jams, 139-143
  Blackberry Jam, 139
  Blueberry Spice Jam, 140
  Fig Jam, 140
  Grape Jam, 140
  Peach Jam, 141
  Pear Jam, 141
  Pear Apple Jam, 142
  Plum Jam, 142
  Strawberry Jam, 142
Jars
  Advantages, 24-25
  Breakage, 23-24
  Canning, 21-25
  Cleaning, 22
  Cooling, 34
  Drying, 222
  Examining, 22-23
  Freezing, 171
  Jelly, 126
  Sterilizing, 22
  Testing Seals, 34
Jellied Spreads, 138-162
  Conserves, 152-155
  Fruit Butters, 155-160
  Fruit Syrup, 161
  Jams, 139-143
  Marmalades, 148-152

Preserves, 143-148
Refrigerated Spreads, 162
Tips for Making, 138-139
Jellies, 132-138
  Apple Jelly, 132
  Blackberry Jelly, 133
  Blueberry Jelly, 133
  Crab Apple Jelly, 134
  Grape Jelly, 134
  Grape Plum Jelly, 135
  Jalapeno Pepper Jelly, 137
  Loquat Jelly, 135
  Mayhaw Jelly, 135
  Mint Jelly, 136
  Muscadine Jelly, 136
  Peach Jelly, 137
  Plum Jelly, 137
  Strawberry Rhubarb Jelly, 138
Jelly, 124-138
  Bag, 127
  Ingredients, 124-125
  Materials and Equipment, 130
  Problems, 130-132
  Recipes, 132-138
  Testing for Doneness, 128-129
  Testing Juice, 127-128
Jelmeter Test, 128
Juice
  Extracting, 126-127
  Testing, 127-128
Juices, canning
  Apple Juice, 42
  Berry Juice, 45
  Grape Juice, 48
  Tomato Juice, 60
  Tomato Vegetable Juice, 60

K

Ketchup, canning
  Blender Ketchup, 68
  Country Western Ketchup, 67
  Tomato Ketchup, 66
Kumquat Marmalade, 150

L

Lamb (also see Meats)
  Canning, 88-89
  Freezing, 203-205
Leathers, Fruit and Vegetable, 237-241
  General Steps, 238
  Puree Recipes, 239-241
  Spicing Up, 238
Suitability of Foods, 237
Lemon Gelatin Glaze, 206
Lemon Verbena, drying, 243

Lemons
  Citrus Fruit Marmalade, 149
  Drying peel, 231
  Green Tomato Marmalade, 150
  Orange Lemon Marmalade, 151
Lids, 25-26
  Testing, 34
Lobsters, in Shell, freezing, 207
Loquat Jelly, 135
Low-Acid Foods
  Canning, 17-18, 59, 87
  Glossary, 8
  Making Jelly, 125
  Table 2, 11
Low-Methoxyl Pectin, 125
Low-Temperature Pasteurization, 97
  Bread and Butter Pickles, 103
  Bread and Butter Zucchini Pickles, 118
  14-Day Sweet Pickles, 104
  Quick Fresh-Pack Dill Pickles, 102
  Quick Sweet Pickles, 105
  Sweet Gherkin Pickles, 106

M

Marinated Hot Peppers, 116
Marinated Whole Mushrooms, 114
Marjoram, drying, 242
Marmalades, 148-152
  Amber Marmalade, 148
  Carrot Orange Marmalade, 149
  Citrus Fruit Marmalade, 149
  Grape Marmalade, 149
  Green Tomato Marmalade, 150
  Kumquat Marmalade, 150
  Orange Lemon Marmalade, 151
  Pear Marmalade, 151
  Pear Honey Marmalade, 152
  Persimmon Marmalade, 152
Materials
  Canning, 21-30
  Drying, 219-222
  Freezing, 170-175
  Making Jelly, 130
  Making Pickles, 98
Mayhaw Jelly, 135
Measuring Equivalents, 15
Meatless Spaghetti Sauce, 68
Meats (also see Poultry)
  Aging, 204
  Broth, canning, 89
  Canning, 87-89
  Chilling, 204
  Chops, canning, 88
  Cutting, 204
  Freezing, 204

Ground, canning, 88
Roasts, canning, 88
Slaughtering, 204
Steaks, canning, 88
Table 7, 14
Thawing, 205
Wrapping, 204
Meats, Cooked, freezing, 214
Melons
  Freezing, 190
  Watermelon Rind Pickles, 121
Methods
  Canning, 30-35
  Drying, 222-230
  Freezing, 175-181
  Making Pickles, 96-98
Mexican Tomato Sauce, 65
Microorganisms
  Canning, 16
  Freezing, 169
  Glossary, 8
Milk, freezing, 211
Mincemeat Pie Filling, canning, 57
Mint, drying, 243
Mint Jelly, 136
Mixed Vegetables, canning, 84
  Pickled Mixed Vegetables, 118
  Vegetable Soup, canning, 85
Modified Pectin, 125
Mold, 8
Muscadines
  Muscadine Jelly, 136
  Muscadine Preserves, 143
  Pear Muscadine Conserve, 155
Mushrooms
  Canning, 78
  Drying, 234
  Marinated Whole Mushrooms, 114
Mustard Greens
  Canning, 77
  Freezing, 199
Mycotoxins, 8

N

Native Persimmon Sauce, 52
Nectarines
  Canning, 49
  Drying, 231
  Freezing, 190
Nuts
  Canning, 93
  Drying, 244
  Freezing, 210

O

Okra
  Canning, 78
  Drying, 235
  Freezing, 200
  Mealed, freezing, 200
  Pickled Dilled Okra, 114
  Tomatoes With Okra or Zucchini, canning, 63
Onions
  Canning, 79
  Drying, 235
  Freezing, 200
  Garden Relish, 110
  Onion Relish, 111
  Pickled Pepper Onion Relish, 108
Oranges
  Amber Marmalade, 148
  Canning, 50
  Carrot Orange Marmalade, 149
  Citrus Fruit Marmalade, 149
  Drying, 231
  Freezing, 191
  Orange Lemon Marmalade, 151
Oven Drying, 220
Oysters, shucked
  Canning, 92
  Freezing, 208

P

Packaging Materials
  Canning, 21-22
  Drying, 222
  Freezing, 171-172
Packing Methods
  Canning, 30-31
  Drying, 228-229
  Freezing, 175-179
    Drugstore Wrap, 176-177
    Butcher Wrap, 178
Parsley, drying, 235, 242
Pasteurization, Low-Temperature, 97
Pasteurizing
  Drying, 228
  Glossary, 9
  Making Pickles, 97
Peaches
  Canning, 51
  Drying, 232
  Freezing, 191
  Peach Butter, 157
  Peach Jam, 141
  Peach Jelly, 137

Peach Pickles, 120
Peach Pineapple Butter, 158
Peach Preserves, 144
Peach Pie Filling, canning, 58
Fresh Peach Puree, 240
Peanuts
  Boiled, in Shell
    Drying, 244
    Freezing, 210
  Green, in Shell
    Canning, 93
    Drying, 244
    Freezing, 210
  Roasted, drying, 244
  Shelled, freezing, 210
Pears
  Canning, 51
  Drying, 232
  Freezing, 191
  Gingered Pear Preserves, 145
  Pear Apple Jam, 142
  Pear Butter, 158
  Pear Conserve, 155
  Pear Honey Marmalade, 152
  Pear Jam, 141
  Pear Marmalade, 151
  Pear Muscadine Conserve, 155
  Pear Pickles, 120
  Pear Preserves, 144
  Pear Relish, 111
Peas
  Dried, canning, 73
  Garden (English, Green)
    Canning, 79
    Drying, 235
    Freezing, 200
  Southern (Blackeye, Crowder, Field)
    Canning, 80
    Freezing, 200
Pecans, freezing, 210
Pectin, 124-125
  Low-methoxyl, 125
  Modified, 125
Pectin Syrup, 185
Pepper Sauce, 116
Peppers
  Hot (Banana, Chili, Hungarian, Jalapeno)
    Canning, 80
    Drying, 235
    Freezing, 201
  Jalapeno Pepper Jelly, 137
  Marinated Hot Peppers, 115
  Pickled Hot Peppers, 115

Pimiento
  Canning, 81
  Drying, 235
  Freezing, 201
Sweet (Bell, Red, Yellow)
  Canning, 81
  Corn Relish, 109
  Dixie Relish, 110
  Drying, 235
  Freezing, 201
  Garden Relish, 110
  Pickled Bell Peppers, 117
  Pickled Pepper Onion Relish, 108
Persimmons
  Drying, 232
  Japanese
    Sauce, canning, 52
    Freezing, 192
  Native
    Sauce, canning, 52
    Freezing, 192
  Persimmon Butter, 159
  Persimmon Marmalade, 152
  Persimmon Puree, 240
pH
  Glossary, 9
  Table 2, 11
Piccalilli, 108
Pickle Relish (see Relishes)
Pickled Eggs, 122-123
  Dark and Spicy Eggs, 123
  Dilled Eggs, 123
  Red Beet Eggs, 123
  Rosy Pickled Eggs, 123
  Spicy Eggs, 123
  Sweet and Sour Eggs, 123
Pickled Fruits
  Peach Pickles, 120
  Pear Pickles, 120
  Spiced Apple Rings, 119
  Spiced Crab Apples, 119
  Watermelon Rind Pickles, 121
Pickled Vegetables
  Bread and Butter Zucchini Pickles, 118
  Corn Relish, 109
  Dixie Relish, 110
  Garden Relish, 110
  Marinated Hot Peppers, 116
  Marinated Whole Mushrooms, 114
  Onion Relish, 111
  Piccalilli, 108
  Pickled Beets, 113
  Pickled Bell Peppers, 117
  Pickled Cauliflower, 113

Pickled Dilled Beans, 112
Pickled Dilled Okra, 114
Pickled Green Tomato Relish, 109
Pickled Hot Peppers, 115
Pickled Mixed Vegetables, 118
Pickled Pepper Onion Relish, 108
Pickled Sweet Green Tomatoes, 117
Pickled Three-Bean Salad, 112
Sauerkraut, 100
Pickles, Making, 94-123
  Alum, 96
  Builders Lime, 96
  Crock, Stoneware, 98
  Fermentation, 97-98
  Fermented Pickles, 94
  Firming Agents, 96
  Fresh-Pack Pickles, 94
  Glossary, 9
  Ingredients, 94-96
  Materials and Equipment, 98
  Methods and Procedures, 96-98
  Pasteurization, Low-Temperature, 97
  Problems, 99-100
  Recipes, 100-123
    Bread and Butter Pickles, 103
    Dill Pickles, 101
    14-Day Sweet Pickles, 104
    Pickle Relish, 107
    Quick Fresh-Pack Dill Pickles, 102
    Quick Sweet Pickles, 105
    Refrigerator Dills, 102
    Reduced-Sodium Sliced Dill Pickles, 103
    Reduced-Sodium Sliced Sweet Pickles, 107
    Sweet Gherkin Pickles, 106
Pickling Salt, 95
Pie Filling, canning, 54-58
  Apple Pie Filling, 54
  Blueberry Pie Filling, 55
  Cherry Pie Filling, 56
  Green Tomato Pie Filling, 57
  Mincemeat Pie Filling, 57
  Peach Pie Filling, 58
Pineapples
  Canning, 52
  Drying, 232
  Peach Pineapple Butter, 158
Pinto Beans, freezing, 197
Plums
  Canning, 53
  Damson Plum Conserve, 153
  Drying, 232
  Freezing, 192

Grape Plum Jelly, 135
Plum Butter, 159
Plum Jam, 142
Plum Jelly, 137
Plum Preserves, 145
Plum Sauce, freezing, 192
Popcorn, drying, 244
Pork
  Canning, 88-89
  Freezing, 203-205
Potatoes (Irish, White)
(also see Sweet Potatoes)
  Canning, 82
  Drying, 236
  Freezing, 202
Poultry (Chicken, Game Birds, Turkey)
  Broth, canning, 90
  Canning, 89-90
  Freezing, 205-206
  Packaging, 205
  Thawing, 206
Poultry, Cooked, freezing, 214-215
Preserves, 143-148
  Fig Preserves, 143
  Gingered Pear Preserves, 145
  Muscadine Preserves, 143
  Peach Preserves, 144
  Pear Preserves, 144
  Plum Preserves, 145
  Pumpkin Preserves, 146
  Quince Preserves, 146
  Strawberry Fig Preserves, 147
  Strawberry Preserves, 146
  Watermelon Rind Preserves, 147
Pressure Canners
  Canning, 28-30
  Glossary, 9
  Using the Canner, 32-34
Pretreating
  Drying, 224-225
  Freezing, 183-184
Problems
  Canning, 35-38
  Freezing, 181-182
  Making Jelly, 130-132
  Making Pickles, 99-100
Procedures
  Canning, 30-35
  Drying, 222-230
  Freezing, 175-181
  Making Pickles, 96-98
Pumpkin
  Canning, 82
  Drying, 236

Index

Freezing, 202
Pumpkin Preserves, 146
Pumpkin Puree, Leather, 240
Pumpkin Seeds
  Dried, 245
  Roasted, 245
Purees
  Banana, freezing, 187
  For Leathers
    Canned Fruit, 239
    Fresh Fruit, 239
    Fresh Peach, 240
    Frozen Fruit, 239
    Frozen Strawberry, 240
    Mixed Vegetable, 241
    Persimmon, 240
    Pumpkin, 240
    Spiced Apple, 239
    Tomato, 241
  Fruits, canning, 48
  Vegetables, canning, 71

Q

Quality, Food
  Canning, 18-19
  Drying, 218
  Freezing, 168-169
Quick Fresh-Pack Dill Pickles, 102
Quick Sweet Pickles, 105
Quince
  Quince Butter, 160
  Quince Preserves, 146

R

Rabbit, canning, 89
Raw Pack
  Canning, 30-31
  Glossary, 9
Reconstituting Dried Foods, 229-230
Red Beet Eggs, 123
Reduced-Sodium Sliced Dill Pickles, 103
Reduced-Sodium Sliced Sweet
  Pickles, 107
Refreezing Foods, 179-181
Refrigerated Spreads, 162
  Refrigerated Apple Spread, 162
  Refrigerated Grape Spread, 162
Refrigerator Dills, 102
Refrigerator-Freezer Test, Jelly, 129
Relishes, 107-111
  Corn Relish, 109
  Dixie Relish, 110
  Garden Relish, 110
  Onion Relish, 111
  Pear Relish, 111

Piccalilli, 108
Pickle Relish, 107
Pickled Green Tomato Relish, 109
Pickled Horseradish Sauce, 111
Pickled Pepper Onion Relish, 108
Reprocessing Foods, 34-35
Rhubarb
  Canning, 53
  Freezing, 193
  Strawberry Rhubarb Jelly, 138
Roe, Fish, freezing, 207
Rosemary, drying, 242, 243
Rosy Pickled Eggs, 123

S

Safety, Food
  Canning, 17-18
  Drying, 218
  Freezing, 168-169
Sage, drying, 242
Salads, freezing, 215
Saline Solution, drying, 225
Salsa, Chili, canning, 70
Salt
  Canning, 71
  Pickling, 95
Sandwiches, freezing, 215
Sassafras, drying, 243
Sauces
  Apple, canning, 43
  Chili Salsa, canning, 70
  Fruit Puree, canning, 48
  Mexican Tomato, canning, 65
  Persimmon, Japanese, canning, 52
  Persimmon, Native, canning, 52
  Pickled Horseradish Sauce, 111
  Plum Sauce, freezing, 192
  Seasoned Tomato, canning, 64
  Spaghetti, Meatless, canning, 68
  Spaghetti, With Meat, canning, 69
  Standard Tomato, canning, 64
Sauerkraut, 100
Savory, drying, 242
Scallops, Shucked, freezing, 208
Scoring, Fairs and Exhibits, 163
Seafood
  Canning, 90-92
  Freezing, 206-208
Seasoned Tomato Sauce, 64
Seeds
  Pumpkin
    Dried, 245
    Roasted, 245

Sunflower
  Dried, 245
  Fried, 245
  Roasted, 245
Sheet or Spoon Test, Jelly, 128-129
Shrimp, freezing
  Cooked Shrimp, 208
  Peeled Shrimp, 208
  Shrimp in Shell, 208
Smoked Fish, canning, 92
Snap Beans (see Beans)
Solar Drying, 220
Solution, Ascorbic Acid, 39
Soups
  Canning, 85
  Freezing, 215
Southern Peas, canning, 80
Soy Beans
  Canning, 74
  Freezing, 197
Spaghetti Sauce With Meat, 69
Spaghetti Squash (see Squash)
Spice Bag, 9
Spiced Apple Rings, 119
Spiced Crab Apples, 119
Spicing Up Leathers, 238
Spicy Eggs, 123
Spinach
  Canning, 77
  Freezing, 199
Spoon Test, Jelly, 128-129
Spreads, Jellied, 138-162
  Conserves, 152-155
  Fruit Butters, 155-160
  Fruit Syrup, 161
  Jams, 139-143
  Marmalades, 148-152
  Preserves, 143-148
  Refrigerated Spreads, 162
  Tips for Making, 138-139
Squash
  Spaghetti, freezing, 202
  Summer
    Drying, 236
    Freezing, 202
  Winter
    Canning, 83
Squirrel, canning, 89
Standard Canning Jars, 21-25
Standard Tomato Sauce, 64
Steam Blanching
  Drying, 224
  Freezing, 175
Stewed Tomatoes, canning, 63

Storing Foods
  Canning, 35
  Drying, 229
  Freezing, 169
Strawberries
  Freezing, 193
  Frozen Strawberry Puree, 240
  Strawberry Fig Preserves, 147
  Strawberry Jam, 142
  Strawberry Preserves, 146
  Strawberry Rhubarb Jelly, 138
Style of Pack
  Canning, 30-31
  Freezing, 184-185, 196
  Glossary, 9
Succotash, 83
Sugar Pack, 184
Sugar Substitutes, 185-186
Sulfite Dip, 225
Summer Squash (see Squash)
Sun Drying, 220
Sunflower Seeds
  Dried, 245
  Fried, 245
  Roasted, 245
Sweet and Sour Eggs, 123
Sweet Gherkin Pickles, 106
Sweet Peppers (see Peppers)
Sweet Potatoes
  Canning, 84
  Sliced, freezing, 202
  Mashed, freezing, 203
Syrup Blanching, 225
Syrup Pack, 184-185
Syrup, Pectin, freezing, 185
Syrups
  Berry Syrup, 161
  For Canning, 40

T

Tartaric Acid Crystals, 49, 127, 130
Tartrate Crystals (see Tartaric Acid Crystals)
Temperature Test, Jelly, 128
Testing for Acid, Jelly, 128
Testing for Doneness, Jelly, 128-129
  Refrigerator-Freezer Test, 129
  Sheet or Spoon Test, 128-129
  Temperature Test, 128
Testing for Dryness, 227-228
Testing Jar Seals, 34
Testing Juice for Acid/Pectin, 127-128
  Alcohol Test, 128
  Cooking Test, 127

Jelmeter Test, 128
Testing for Acid, 128
Thawing Eggs, 209
Thawing Fruits, 186
Thawing Meats, 205
Thawing Vegetables, 196
Thyme, drying, 243
Tips for Making Jellied Spreads, 138-139
Tomatoes (also see Green Tomatoes)
  Adding Acid, canning, 59
  Blender Ketchup, canning, 68
  Canning, 59-70
  Chili Salsa, canning, 70
  Country Western Ketchup, canning, 67
  Crushed, canning, 62
  Dried Beans, canning, 74
  Drying, 236
  Freezing, 203
  Garden Relish, 110
  Meatless Spaghetti Sauce, canning, 68
  Mexican Tomato Sauce, canning, 65
  Mixed Vegetables, canning, 84
  Piccalilli, 108
  Seasoned Tomato Sauce, canning, 64
  Spaghetti Sauce With Meat, canning, 69
  Standard Tomato Sauce, canning, 64
  Stewed Tomatoes, canning, 63
  Tomato Halves, canning, 61
  Tomato Juice, canning, 60
  Tomato Ketchup, canning, 66
  Tomato Paste, canning, 66
  Tomato Puree, 241
  Tomato Vegetable Juice, canning, 60
  Whole Tomatoes, canning, 61
  Tomatoes With Okra or Zucchini, canning, 63
Tray Pack
  Freezing Fruits, 184
  Freezing Vegetables, 196
Trays, drying, 221
Turkey (also see Poultry)
  Canning, 89-90
  Freezing, 205-206
Turnip Greens, canning, 77
Turnip Roots, canning, 84

U

Unsweetened Liquid Pack, 185-186

V

Vacuum, 9
Veal
  Canning, 88-89
  Freezing, 203-205
Vegetable Soup, canning, 85

Vegetables
  Canning, 71-86
  Drying, 233-236
  For Leathers, 241
  Freezing, 194-203
  Making Pickles, 112-118
  Mixed Vegetables
    Canning, 84
    Pickled Mixed Vegetables, 118
    Puree for Leather, 241
    Succotash, canning, 83
    Vegetable Soup, canning, 85
Vegetables, Cooked, freezing, 216
Venison
  Canning, 88-89
  Freezing, 203-205
Verbena, Lemon, drying, 243
Vine Drying, 221

W

Water Bath
  Canners, 28
  Using the Canner, 32
Water Blanching, 224
Water Pack, 206
Watermelon Rind
  Watermelon Rind Pickles, 121
  Watermelon Rind Preserves, 147
Wax Beans (see Beans, Snap)
Whipping Cream, freezing, 211
White Potatoes (see Potatoes)
Wild Game (see Meats)
Winning, Fairs and Exhibits, 163-167
Winter Squash (see Squash)
Whole-Kernel Corn
  Canning, 77
  Freezing, 199

Y

Yeasts, 9

Z

Zucchini
  Bread and Butter Zucchini Pickles, 118
  Cubed, Shredded, canning, 86
  Mixed Vegetables, canning, 84
  Tomatoes With Okra or Zucchini, canning, 63